ACTION
SCIENCE

This book is dedicated to Gordon H. Robertson Jr. and Richard E. Ellis—men of great personal impact who supported me, mentored me, and challenged me to set paths to achieving high goals in my education, my work, my activities, and my life.

ACTION SCIENCE

Relevant Teaching and Active Learning

William H. Robertson

CORWIN
A SAGE Company

CORWIN
A SAGE Company

FOR INFORMATION:

Corwin

A SAGE Company

2455 Teller Road

Thousand Oaks, California 91320

(800) 233-9936

www.corwin.com

SAGE Publications Ltd.

1 Oliver's Yard

55 City Road

London EC1Y 1SP

United Kingdom

SAGE Publications India Pvt. Ltd.

B 1/I 1 Mohan Cooperative Industrial Area

Mathura Road, New Delhi 110 044

India

SAGE Publications Asia-Pacific Pte. Ltd.

3 Church Street

#10-04 Samsung Hub

Singapore 049483

Acquisitions Editor: Robin Najar

Associate Editor: Desirée A. Bartlett

Editorial Assistant: Ariel Price

Project Editor: Veronica Stapleton Hooper

Copy Editor: Megan Granger

Typesetter: C&M Digitals (P) Ltd.

Proofreader: Sarah J. Duffy

Indexer: Jean Casalegno

Cover Designer: Edgar Abarca

Marketing Strategist: Maura Sullivan

Photos by Erik Hilburn, E. P. Robertson, Inc.

ISBN 9781452256566

14 15 16 17 18 10 9 8 7 6 5 4 3 2 1

Contents

Acknowledgments

The efforts associated with this work come from a number of people who have contributed to the development and success of the concept of action science in both the areas of action sports and education.

I am indebted to the contributions of the many amazing athletes and performers in skateboarding and BMX who have put their skills and abilities to the test in demonstrations, videos, and photographs. Primarily, I would like to thank Billy Gawrych, an accomplished BMX flatlander and announcer who has dedicated much of his time and effort to bring his talents as both a rider and an artist to the development of action science. He was also instrumental in assembling teams of top professional and amateur athletes who have participated in activities with me over a good many years. Without his spirit, his efforts, and his commitment, this work would not have matured and grown so quickly.

As for the athletes, I have to thank a number of professional and top amateur skateboarders and BMX riders who have been a part of the Action Science Demo Team over the years, including John Andrus, Chris Benker, Fabiola Da Silva, Rayce Davis-Eisenhart, Daniel Dhers, Lewis Dinsdale, Ommar Estrada, Jeff Ferris, Billy Gawrych, Nick Happel, Gary Laurent, Adrian Lenardic, Garrett Morrison, Vic Murphy, John Parker, Brian Saresmaslani, Chauntae Schwoegler, Tom Stober, Art Thomason, Eddie Vargas, Dave Voelker, Morgan Wade, Jacob Whitt, and Paul Zimmerman.

I would like to thank the faculty, staff, and students at the University of Texas at El Paso (UTEP) for the support as well as for the opportunity to integrate action science in my work as a teacher and researcher.

I would also like to acknowledge a group of El Paso, Texas, educators whose efforts and faith in this approach were fundamental in developing video content and classroom activities. Many thanks go to Ashton Graham, Mary Beth Harper, Tim Holt, and Steve Putnicki for their hours of commitment to helping young people achieve success in education through action science. I also have to single out Juliette Caire, who has been an ardent supporter of action science in the great number of educational projects she has done in El Paso and at UTEP. Without her faith and willingness to try new things, much of the ground that has been covered educationally would remain untraveled.

I would also like to thank my wife, Sarah Robertson, who has been an active partner and supporter in all things education and skateboarding in my adult life. She continues to be my inspiration, as well as my grounding force, and I am indebted to her in all areas of my life.

PUBLISHER'S ACKNOWLEDGMENTS

Corwin wishes to acknowledge the following peer reviewers for their editorial insight and guidance.

Susanne L. Hokkanen, seventh-grade science teacher
Colin Powell Middle School
Matteson, IL

Stacy M. Holland, science teacher
Department Chair, Katy ISD
Katy, TX

Melissa Miller, MS science teacher
Lynch Middle School
Farmington, AR

About the Author

Dr. William H. Robertson has been an educator for more than 20 years and has taught at the middle school, high school, and university levels. His academic areas of expertise are in science education, curriculum development, and technology integration for K–12. Additionally, he develops, researches, and teaches materials related to problem-based learning and action science.

He completed his PhD in multicultural teacher and childhood education, with an emphasis in science and technology, and has taught hundreds of pre-service and in-service teachers in the areas of science and technology in classrooms, workshops, and trainings in the United States, Canada, Mexico, and throughout South America. He also holds a master's degree in science education from the University of Colorado–Boulder, a BA in Spanish from the University of Texas at El Paso, a BS in biology from Northern Arizona University, and a BA in history from Duke University.

As a longtime participant and performer in skateboarding, with more than 35 years of experience in the sport, he has developed Dr. Skateboard's Action Science (http://www.drskateboard.com), which addresses physical science concepts for middle school students, utilizing skateboarding and bicycle motocross (BMX).

He is uniquely qualified to exhibit excellence in science teaching and curriculum development combined with the integration of action sports such as skateboarding and BMX, and this method and combination have been thoroughly tested and presented, achieving great results both nationally and internationally.

Forces are used to change the motion of an object, and in skateboarding and BMX, the rider's ability to control and affect these forces is what allows for cool tricks and ripping moves. Paul Zimmerman, cameraman. Shown: William H. Robertson.

1   What Is Action Science?

It's what you learn after you know it all that counts.

—John Wooden

Action Science: Relevant Teaching and Active Learning explores concepts suitable for middle-grade students, set in a relevant and relatable curriculum designed to address both the objectives and enduring knowledge of physical science in content and process skills for the Next Generation Science Standards. The entire *Action Science* curriculum maps to fundamental content in physical science with which middle-grade students should engage. The activities and methods associated with *Action Science* include content for the classroom teacher, activities to be used in the classroom, and engaging video content that focuses on physical science concepts related to motion, forces, Newton's Laws of Motion, and simple machines. A fundamental purpose of this approach is to link the defined concepts of physical science to skateboarding and bicycle motocross (BMX), and to provide an interesting method of engaging students in the exploration of science in a real-world context. The overarching theme for *Action Science* is that activities such as skateboarding and BMX can be practical means for helping students integrate their personal interests and youth culture with teaching and learning inside and outside of the classroom.

I work as a university professor in the areas of science education, curriculum development, and technology integration. As an educator and a skateboarder, I knew I would have unique opportunities to teach and provide new methods to instruct students and aid teachers. With a PhD in science and technology education, I am able to apply solid instructional principles to the courses I teach at the university level and, on a broader scale, in the community through outreach activities that feature skateboarding.

Through skateboarding, I personally have learned patience, discipline, creativity, and the art and science of practice. My audiences for on-site demonstrations have included elementary, middle, and high school students in academic settings around the country. I have gained much of my experience in schools, and now, with *Action Science,* schools can pursue new options to add relevant educational materials for middle-grade students. Utilizing this platform to teach the physics of skateboarding has given me the unique identity of Dr. Skateboard. The Dr. Skateboard website (www.drskateboard.com) provides details on performances, as well as curriculum materials for students and families. My audiences of children and parents typically do not see the connection between skateboarding and physics. They wonder, "If you have a PhD, why do you skateboard?" The answer is simply that skateboarding is fun and a part of who I am.

Action Science was developed with the help of a collaborative team that included university faculty, school district personnel, middle school teachers, and local students. Additionally, video scripts were written and then used as a guide for capturing live footage at a series of on-site school demonstrations, as well as for developing a context for delivering science content in settings such as skate parks, classrooms, and local community sites.

In addition, the Action Science Demo Team, a group of professional athletes specializing in skateboarding and BMX, demonstrated the action scenes that appear in the videos and photos that accompany this text. With their extensive experience delivering performances in educational settings, this group lends authenticity to the video segments, and their athletic abilities serve as an important link to the science content being presented. These athletes perform high-flying maneuvers that demonstrate physical science concepts, such as the relationship between velocity and acceleration. Without the athletes, the sport's scientific relevance would not be as apparent. This is another pathway that invites learners to learn, in that they may not initially be attracted to science but may recognize and respect the difficulty of the maneuvers performed in each video segment. The videos provide participating teachers and students with a series of engaging instructional opportunities, embedded in classroom activities that provide relevant content information that can be used to explore and explain scientific concepts and to engage students in active learning.

This book is grounded in the implementation of a constructivist teaching and learning methodology that centers on a teacher first providing students with an experience and then sharing the content of what they will learn after completing an activity, discussing ideas, and being immersed in the content in a hands-on manner. Chapters 2, 3, and 4 detail a path that links the 5Es of Constructivism to the practical methods of teaching in the classroom. Chapters 5, 6, and 7 follow that foundation of constructivism and put forward a way of using action science in classroom situations with resources that can be activated pragmatically.

The classroom activities (Chapter 8) were developed in part to allow teachers the opportunity to use readily available common household materials in the context of experiments that help the learner explore and explain the concepts presented in the video instruction. The provided activities were developed and field-tested by middle school science teachers, whose feedback helped shape the design, development, and implementation of the classroom activities and video instruction. These resources are provided to help the classroom teacher appeal to students' interests and create ways to further integrate topics of youth culture into the curriculum. The goal is to provide topics of relevance that are initially engaging, using hands-on and immersive experiences to tap into student motivation that will lead to increases in content.

When a rider goes into a 360 air, the centripetal force pushes in on the rider. The experienced rider compensates for this force by using his or her body to balance the force and continue in a spinning motion. Shown: John Andrus.

Action Science provides a resource for teachers to engage students in self-directed learning, participating in collaborative teams, acquiring critical knowledge, and developing proficiency in problem solving. For the educator, the video segments and activities should be used in tandem, as the videos provide context, serve as engaging concept introductions, and also contribute relevant content to each lesson. The video segments may be shown in their entirety within a given episode, or the instructor may choose to use a portion of the video to highlight a given topic to be explored in a classroom activity. For example, prior to introducing the activity "Flatland BMX and the Center of Gravity" (see Activity 1 in Chapter 8), the teacher may want to show the video segment on the center of gravity. For each of the activities, the provided video segments can be used as hooks to introduce the activity and as a final content review once the activity has been completed.

As teachers implement the *Action Science* classroom activities and video materials, the fundamental goal is to positively impact student success in physical science content knowledge. The classroom activities and video instruction map directly to the physical science standard objectives. The supporting chapters provide rich context and direction for the classroom teacher in choosing a method and teaching approach for delivering these materials. The videos and classroom activities are also important to provide evidence of students' conceptual understandings of science topics and to gauge changes in how teachers and students develop and explain their understandings in the given physical science content area. Although this is not to claim that a single method or activity can uniquely impact student achievement, the hope is that as teachers use this unique approach within the curriculum materials, they will see gains in student learning.

By immersing students in an approach for learning science that is based on action sports and focused on physical science, the premise is that students' process skills and overall content knowledge will increase. The purpose of *Action Science* is to positively impact middle school students' achievements in physical science, as well as to provide teachers with the resources to engage students in meaningful, relevant science content. The long-term prospect of this approach is to demonstrate that the implementation of a curriculum approach built around student interests such as skateboarding and BMX can positively impact student achievement in science content and conceptual understanding. Additionally, teachers using this approach have reported increased interest and focus among previously marginalized students.

TRANSFORMATIVE SCIENCE

Action science is an example of the use of transformative educational strategies to enhance the study of science for middle school students. *Action science* can be defined as the use of familiar objects, circumstances, and situations within students' lives to explain specific concepts in science built around student interests, including action sports such as skateboarding and BMX. In schools, these topics are often approached in more traditional ways that employ content delivery mechanisms not put in relevant terms for the middle school learner.

The process by which students move from acquiring factual knowledge to changing in meaningful ways as a result of what they have learned is known as transformative education. Although transformative education has largely been associated with adult learning theory, there is support for considering transformative learning theory as a possible medium for working with middle school students. First, helping young adolescents transition into adulthood requires that they develop a deeper understanding of concepts and issues that will lead them to question their fundamental beliefs and assumptions, thus resulting in a transformation of perspective or worldview. Second, preparing middle school students to develop critical and reflective thinking skills encourages them to care about the world around them and to recognize that some degree of personal or social transformation is required.

Currently, the idea of implementing transformative education in middle school classrooms is gaining momentum because of its link with constructivist pedagogy. Constructivism is a popular and successful learning theory that suggests that students actively construct and reconstruct knowledge, thereby transforming meanings to arrive at new understandings and multiple ways of thinking (Brooks & Brooks, 1993). The work of John Dewey, Maria Montessori, Jean Piaget, Jerome Bruner, and Lev Vygotsky is presented in the constructivist view of teaching and learning. Students learn to consider multiple points of view and to question assumptions, values, and beliefs, while always seeking to verify reasoning. In this way, the goal of a classroom teacher implementing a student-centered strategy founded in constructivist methods is to make middle school science transformative—relevant, pragmatic, relatable, approachable—in this case, through the integration of action sports and science content.

For education to be transformative, the traditional relationship between the teacher and the student, wherein the teacher delivers content while the student listens passively, is rejected. Instead, teachers serve as facilitators, co-learners, mentors, and role models. Teachers need to present themselves as respectful guides and compassionate helpers who grant students opportunities to become actively involved in classroom interactions and in their own learning (Hasslen, 2008). More important, teachers become change agents by working to establish links within their communities and by trying to engage their students in active learning projects that require them to interact with individuals outside the school (Donovan, 2002). For the transformative education teacher, learning can take place in different venues and not solely in the classroom (Palmer, 1998).

The use of *Action Science* as a mechanism for integrating transformative education is an approach that appears to be enhancing the interest and motivation of middle school students in science. In addition, *Action Science* contributes to the field of education in that it aims to determine how transformative education can be used to motivate marginalized middle school students to learn science. Studies have shown that students involved in active learning in meaningful contexts acquire knowledge and become proficient in problem solving (Robertson, 2008). If alternative-education students who are often reluctant to become engaged in schoolwork can come to enjoy learning concepts in physics, such as forces and motion, imagine the possibilities for students enrolled in regular education programs. As with all middle school students—but even more so with marginalized students—science education needs to be transformed, and action science is a transformational example of an effective approach toward accomplishing this goal.

Certain forces such as gravity and lift are important in under-standing the basic mechanics of flight. In skateboarding, a rider can boost an air above the ramp by increasing lift and being able to handle accelerated speed. Shown: Rayce Davis.

In the classroom, constructivist curriculum must be designed to reflect real-life situations (Bentley, 1995). The use of students' social context within the curriculum organizes the content and repositions the relevance of scientific concepts (Hofstein & Yager, 1982). In the real world, research scientists seldom operate exclusively in isolated areas of content; they cross over the barriers between disciplines all the time and integrate content knowledge and process application. Immersive and practical experiences give students the ability to retain facts by logically working through problems that require thinking and application. As Bruner (1962) stated, "Students should know what it feels like to be completely absorbed in a problem. They seldom experience this feeling in school" (p. 50).

Students exploring concepts should work with familiar materials in hands-on activities so they can experience situations that are real and fundamental. Student-centered learning plays a valuable role in the constructivist paradigm, as it is the process of learning by doing (Dewey, 1902), utilized fundamentally in the phases of constructivism. For example, teachers with students who enjoy skateboarding can provide opportunities in the classroom to explore the concepts of velocity, acceleration, center of gravity, and moment of inertia. They may also use the skateboard and a local skate park to investigate topics such as inclined planes, levers, fulcrums, and screws. The purpose of this approach is to allow students to explore meaningful science topics in the context of an activity they enjoy. Learners have so much fascinating content at their fingertips, and with technology becoming more affordable, they can access more and more of this content from their homes. It is important to engage learners in situations that effectively integrate their own experiences and familiar materials they can use to better understand specific concepts (Eisenkraft, 2003).

The learner is always defining meaning within the context of action and reflection, as meaning is a human construction within a social situation (Brooks & Brooks, 1993). Students need opportunities to address misconceptions and develop concepts in real-world situations. Yet educators must beware of regarding the learner's point of view as fully complete and significant in and of itself (Dewey, 1902). Each learner understands content and concepts differently based on his or her previous experiences. "Students come to school with their own ideas, some correct and some not, about almost every topic they are likely to encounter" (Rutherford & Ahlgren, 1990, p. 198). Learning is the responsibility of the learner, and the teacher must guide students toward developing understandings from content materials and classroom experiences.

How can an object stay in motion? Well, inertia is the resistance of an object to a change in speed or direction, and with inertia, objects stay in motion. Shown: Art Thomason.

As students explore content in active and relevant ways, they develop a broader understanding of the concepts in application. When they collaboratively relate what they are learning, seeing, or doing, they can begin to see similarities in their understandings, as well as self-identify misconceptions they may have about content material (Bybee et al., 2006). This sharing within cooperative groups is a fundamental strategy in constructivism as it allows the teacher to facilitate the learning process and also helps develop a common base of experiences through which students can connect to the content. Hands-on explorations of simple topics that use problem-solving strategies, combined with collaborative interactions among students, help build an understanding of processes and concepts (Apple, 1993).

The use of action science as a mechanism for integrating transformative education is an approach that appears to be enhancing the interest and motivation of middle school students in science. It is the purpose of *Action Science* to positively impact achievement for middle school students in physical science knowledge and skills. By immersing students in a science learning approach based on action sports and focused on the goals and objectives of physical science, students' process skills and overall content knowledge have the potential to greatly increase.

REFERENCES

Apple, M. W. (1993). *Official knowledge.* New York, NY: Routledge.

Bentley, M. L. (1995). Carpe diem. *Science Activities, 32*(3), 23–30.

Brooks, J. G., & Brooks, M. G. (1993). *In search of understanding: The case for constructivist classrooms.* Alexandria, VA: Association for Supervision and Curriculum Development.

Bruner, J. (1962). *The process of education.* Cambridge, MA: Harvard University Press.

Bybee, R. W., Taylor, J. A., Gardner, A., Van Scotter, P., Powell, J. C., Westbrook, A., & Landes, N. (2006). *The BSCS 5E instructional model: Origins, effectiveness, and applications; executive summary.* Retrieved from http://www.bscs.org/sites/default/files/_legacy/BSCS_5E_Instructional_Model-Executive_Summary_0.pdf

Dewey, J. (1902). *The child and the curriculum.* Chicago, IL: Chicago University Press.

Donovan, B. (2002). An illustration of practice in search of theory. *Theory and Practice, 41,* 17–26.

Eisenkraft, A. (2003). Expanding the 5E model. *The Science Teacher, 70*(6), 57–59.

Hasslen, R. (2008). From a tarpaper shack to a transformed classroom: A teacher's journey. *Kappa Delta Pi Record, 44,* 52–54.

Hofstein, A., & Yager, R. (1982). Societal issues as organizers for science education in the '80s. *School Science and Mathematics, 82*(7), 539–547.

Palmer, P. J. (1998). *The courage to teach: Exploring the inner landscape of a teacher's life.* San Francisco, CA: Jossey-Bass.

Robertson, W. H. (2008). *Developing problem-based curriculum: Unlocking student success utilizing critical thinking and inquiry.* Des Moines, IA: Kendall Hunt.

Rutherford, F. J., & Ahlgren, A. (1990). *Science for all Americans.* New York, NY: Oxford University Press.

Kinetic energy is energy in motion, and a BMX rider doing a cool trick is certainly an example of this. Shown: Daniel Dhers.

2 Constructivism and the Classroom Teacher

Obstacles are those frightful things you see when you take your eyes off your goal.

—Henry Ford

For education to be constructive, the traditional teacher–student relationship—historically consisting of the teacher delivering content while students listen passively—must be discarded. Instead, teachers should serve as facilitators, mentors, role models, co-inquirers, and friends while helping students seek understanding of the classroom curriculum content. Teachers need to view themselves as respectful guides and compassionate helpers who provide students with opportunities to become actively involved in their own learning and classroom operations. An active learning environment in which the teacher integrates oral, visual, and kinesthetic strategies allows for learning to center on the students. In this manner, teachers become change agents, linking students' relevant life experiences to curriculum content—and in no area is this more needed than middle school science. Teachers must work to establish links within their learning communities and try to engage their students in active learning projects that require them to interact with individuals inside and outside the school. The constructivist science teacher needs to extend learning into the fabric of students' lives, not treat it solely as a subject to be explored uniquely in the classroom.

To enable practical implementation of such an approach, educators should utilize teaching strategies that emphasize providing experiences first and content delivery second. One such method that is valuable as a pedagogical and curriculum organizer is constructivism, which is a learning strategy that builds on students' existing knowledge, beliefs, and skills (Brooks & Brooks, 1993). "It includes skills and activities that increase curiosity for research, satisfy student's expectations, and make the student focus on an active research for information and understanding" (Ergin, Kanli, & Unsal, 2008, p. 57). Within a constructivist approach, as students encounter new information, they work to synthesize understandings based on their current experiences and prior learning. Students assemble meaning while continually self-assessing their understandings of concepts in the context of their own world experiences. In other words, the constructivist approach to learning states that learners of all ages build new ideas on top of their personal conceptual understandings (Eisenkraft, 2003). Through this process, students and teachers experience common activities while applying the requirements outlined in the Next Generation Science Standards, which define the specific science content, practices, and interdisciplinary connections that link to the comprehensive ideas and practices students need to know.

A five-phase process known as the 5Es characterizes constructivism. The 5Es consist of the engagement phase, exploration phase, explanation phase, elaboration phase, and evaluation phase (Bybee, 2003). "The important point is that each (learner) has their own construction, their own understanding, rather than some common reality" (Duffy & Jonassen, 1992, p. 6). With this colearning approach, students and teacher are enabled to construct a deeper and more comprehensive understanding through activities that match their cognitive capabilities and are delivered in a framework that sparks motivation, incites inquiry, and then, as a result of collective experience, delivers content knowledge in conceptually correct contexts. The key to the constructivist method is to build on previous learning and apply new learning in a meaningful context, which centers on active learning and requires learners to address their own understandings in the context of new experiences and learning opportunities.

ENGAGEMENT WITHIN ACTION SCIENCE

Establishing a relevant and relatable connection to content is critical to gaining student interest and increasing motivation in classroom topics, especially in the areas of mathematics and science. For middle school students, a critical point comes at the beginning of a lesson or program, as they quickly decide if they will actively participate or withdraw from instruction.

Momentum is the measure of mass in motion, and this property depends on both mass and velocity. Shown: Jacob Whitt.

Engagement activities should help students make connections between past and present learning experiences, to move them toward thoughtful involvement in the concept, process, or skill to be learned. In other words, the student should relate to the problem being posed and be invested in pursuing a solution. Previous studies using skateboarding—specifically, the construction of ramps—as a hook to engage students in real-world applications of mathematics "lend support to the argument that all students can benefit from and deserve the opportunity to engage with interesting and challenging problems" (Stephens, Bottge, & Rueda, 2009, p. 525).

One example of an engagement strategy would start with using the video segment titled "The Center of Gravity," located on the *Action Science* website. This video presents the concept of the center of mass and bridges the concepts of gravity and lift, which are part of a collection of physical science forces that middle school students need to know. Prior to showing the video segment, the teacher can pose open-ended questions to activate students' previous knowledge concerning this content. Sample questions include "What do you do when you ride a skateboard or bicycle?" "How do you balance on a skateboard or bike?" and "What forces are acting on you as you are trying to ride a bike or skateboard?" Additionally, previously marginalized students who have experience in these activities but may struggle in science can become experts in this discussion and contribute greatly to the classroom investigations.

Finally, the teacher should conclude the series of questions by asking, "What is the center of gravity, and why is it important?" The teacher can then transition the conversation to introduce the video segment, which serves as the engagement for the activity "Flatland BMX and the Center of Gravity." This classroom activity (see Activity 1 in Chapter 8 of this book) is designed to allow students to explore the concept of the center of gravity in the classroom, as students create irregular cardboard shapes and determine the object's center of gravity through a series of step-by-step procedures. Students exploring a concept should be given opportunities to work with familiar materials in a hands-on manner so they can have kinesthetic, visual, and collaborative experiences. Hands-on learning plays a valuable role in the constructivist paradigm, as it is the process of learning by doing (Dewey, 1902) utilized in explorations and experiments. Look at the action science video titled "Dr. Skateboard's Action Science—Forces—Episode 2—Center of Gravity," accessed with the QR code below.

EXPLORATIONS ARE DRIVEN BY STUDENT QUESTIONS

Have you ever gotten a new computer or cell phone? What was the first thing you did? If you are like most people, the first thing you did was turn it on and begin to explore its functions based on what you had learned previously on your own or with the aid of others. For most people, this type of exploration will continue unbridled until some problem is encountered or some aspect requiring further understanding is revealed. Generally, at that moment, a person will consult a manual or get some help from a friend or colleague to better understand exactly what is going on and how best to correct it. In this way, the importance of the exploration is that it provides an experience that builds a foundation for content delivery and understanding for further learning.

A quality exploration activity is central to building on the initial aspects of engaging students. In the case of students working with *Action Science*, specific activities included in this book have been designed to incorporate asking questions, developing teamwork, and gathering data. In a constructivist framework, the exploration phase should provide students with a common base of experiences and build directly on the motivation to learn inspired by engagement activity. As students actively explore their environment, they are already learning, and the teacher can provide an environment for inquiry in which students identify and develop concepts, processes, and skills based on an open-ended approach. "The correlation between the subjects taught in Science and especially Physics lessons and daily life is very important" (Ergin et al., 2008, p. 57). The purpose of this approach built on exploration, asking questions, and seeking answers within the exploration phase is to allow students to explore meaningful science topics within the context of something they enjoy doing.

For example, the teacher may want teams of students to gather data from three different stations in a local skate park, which seems an unlikely place to study science. Of course, this activity would require real athletes to perform specific maneuvers. In this three-station example, the first station could be a half-pipe—a semicircular ramp structure—where riders would travel back and forth while students calculated angular motion. The second station could be an inclined plane about 1 meter tall and 3 meters long. As the riders dropped in on the inclined plane ramp, students could record the time it took each rider to reach the edge of the ramp. From this data, the students could calculate each rider's acceleration. The third station in our example might be a grind bar—a horizontal metal pole affixed to the ground. The student teams would need to calculate the rider's velocity as he or she rode across the grind bar, and determine the minimal velocity needed for a rider to travel the length of the bar.

Objects in motion and objects at rest both have forces acting on them. Those forces can be balanced or unbalanced. What is a balanced force? Well, it is any force that does not cause a change in the direction or speed of an object. Shown: John Andrus.

As students explore such science concepts as those presented here—acceleration, velocity, and angular motion—in a real context, they develop a broader understanding of those principles in the context of their own experience. When students are able to share their collective observations and understandings through small-group discussions, they are able to strengthen their understandings of the scientific concepts. This approach has also been shown to increase congruence in teaching, an instructional strategy that aligns the coherent relevance of the curriculum with the specific content knowledge and skills of a lesson to create optimal learning (Bybee, 2003). This sharing within cooperative groups is a fundamental strategy in the constructivist approach, as it allows the teacher to facilitate the learning process and also helps students develop a common base of experiences on which to make connections to content. The teacher can then best use the knowledge and skills from open-ended, field-based experiences to help students take responsibility for their own learning, which is a fundamental tenet of the constructivist method.

EXPLANATIONS ARE TIMES TO PRESENT PRIMARY CONTENT

Going back to the example of a new cell phone or computer, once a person has explored a problem and cannot fully understand the next steps, there is a need for new content that maps to the experiences found during exploration. In other words, the person is ready for additional content because of the experience in which he or she has engaged, and terms applicable to specific functions or situations take on new meaning, as they are now presented in connection with the learner's previous experience.

Getting students engaged and exploring concepts must invariably help them master content, and this approach should extend beyond purely prescriptive approaches. When students receive authentic tasks that allow them to directly manipulate data, they uncover content relevant to the ideas they have been exploring. In the skate park example, after gathering the data in teams, students would have to make mathematical calculations, discuss their results, and justify their solutions within each group. This strategy requires student teams to actively interact with the content of the lesson, collate the content from any provided worksheets, and discuss their collective experience to provide logical solutions requiring analysis and synthesis of information. In the areas of science and mathematics, often called the "language of science," reasoning and making sense of content in context are critical to helping students organize their knowledge in ways that enhance the development of conceptually correct understandings (Martin, 2009).

ELABORATIONS DEEPEN CONNECTIONS TO CONCEPTS

As students gain experiences in science and then seek understanding through interacting with primary content, the teacher can dive deeper into topics that need clarification. This also provides an opportunity for students to move past memorizing content as facts, instead making the content part of their collective learning so the skills of information analysis and synthesis into new situations are explored and explained. For example, students may choose from a menu of activities, which can center on classroom topics in the action science areas of forces, motion, Newton's Laws of Motion, or simple machines. The purpose is to provide students with a menu of activities based on their interests, as well as hands-on explorations that focus on specific concepts in physical science that align with the national Next Generation Science Standards.

This elaboration phase is designed to extend students' conceptual understanding into applications of skills and behaviors and to deepen and broaden their content knowledge. During activities in the classroom, the students would again be assembled into teams, albeit it in new groups to provide new perspectives and collaborations. The students then would gather data and solve problems in the focus areas that centered on content being covered in the classroom. For example, in the activity on simple machines titled "Skateboards Have Levers and Fulcrums" (see Activity 7 in Chapter 8), the students have to construct a catapult and calculate distances and the angle of release for each trial. In doing so, the students also address specific concepts pertaining to levers and fulcrums and the use of scientific thinking and reasoning. In this manner, *Action Science* can impact science, technology, engineering, and mathematics education by engaging students in a learning process that is integrative and purposeful. This emphasizes the interdisci-

Friction causes a change in inertia, but most riders know this and will use these forces to find a balance between staying in motion and remaining at rest. Shown: Rayce Davis.

plinary nature of science through classroom activities now extended into the elaboration phase, a connection that fosters deeper and broader understandings of science's relation and relevance to real life.

EVALUATION DEMONSTRATES THE PROGRESS

The use of creative learning situations, such as gathering data from real athletes at the skate park, can provide a context for students to ask their own questions about their learning experiences as they develop scientific content knowledge related to physical science. The worksheets, activities, quizzes, and tests can all be part of the teacher's classroom evaluation, and the constructivist method also requires a final demonstration by the students. For example, students may be asked to construct their own catapults using a set of provided materials and then, in three trials, to see who can launch an object such as a marshmallow the farthest. They may also be asked to explain the concepts of the lever and fulcrum as they pertain to their specific catapult design. Within a culminating project such as this, students should demonstrate their understandings, as this type of evaluation provides an opportunity for each student to gauge personal progress and for the teacher to see exactly what students understood as a result of the experiences gained through *Action Science*.

In summary, the evaluation phase requires learners to assess their own understanding and abilities, as well as allowing the teacher to evaluate students' understanding of key concepts and skill development. The assessments should be both formal and informal and continuous so the teacher can best help students learn at their own levels. In the case of *Action Science*, conducting the activities in a constructivist framework gives students a chance to assess their own understanding of concepts required in physical science.

SUMMARY

How do educators get students to enjoy science? *Action Science* proposes to answer this question by filling each day in the classroom with exciting activities that allow students to discover connections between mathematics and skateboarding. As key elements are presented to keep students engaged and ready to learn, the students' inherent motivation to learn is activated and, through the constructivist method, the teacher fosters and facilitates this desire for understanding. If students are "hooked" by an opening activity because they connect to skateboarding or enjoy visiting a local skate park to gather data, the teacher has, in effect, captured their interest and activated their ability to relate effectively to the content.

By engaging, exploring, and explaining the content in relevant terms and experiences, the students can elaborate on their skills and understandings by doing other activities directly connected to their various interests. Finally, students should have many opportunities to evaluate their own conceptual understanding through each day's classroom activities. "Essential to expertise is mastery of concepts that allow for deep understanding. Such understanding helps the learner reformulate facts into useable knowledge" (Bybee, 2003, p. 350).

Classroom teachers often face the pressures of high-stakes testing and covering massive amounts of material in limited periods of time. In science, students are not often engaged and do not seem to enjoy learning skills, nor do they usually see connections between the real world and the topics under study. By implementing the constructivist approach of the 5E model in an innovative and creative way, as presented in *Action Science*, students will be immersed in required content and able to participate at a higher cognitive level in an enjoyable and student-centered manner.

REFERENCES

Brooks, J. G., & Brooks, M. G. (1993). *In search of understanding: The case for constructivist classrooms*. Alexandria, VA: Association for Supervision and Curriculum Development.

Bybee, R. W. (2003). The teaching of science: Content, coherence, and congruence. *Journal of Science Education and Technology, 12*(4), 343–358.

Dewey, J. (1902). *The child and the curriculum*. Chicago, IL: Chicago University Press.

Duffy, T., & Jonassen, D. (1992). *Constructivism and the technology of instruction: A conversation*. Hillsdale, NJ: Lawrence Erlbaum.

Eisenkraft, A. (2003). Expanding the 5E model. *The Science Teacher, 70*(6), 57–59.

Ergin, I., Kanli, U., & Unsal, Y. (2008). An example for the effect of 5E model on the academic success and attitude levels of students': "Inclined projectile motion." *Journal of Turkish Science Education, 5*(3), 47–59.

Martin, G. (2009). *Focus in high school mathematics: Reasoning and sense making*. Reston, VA: National Council of Teachers of Mathematics.

Stephens, A., Bottge, B., & Rueda, E. (2009). Ramping up on fractions. *Mathematics Teaching in the Middle School, 114*(9), 520–526.

Another set of forces skateboarders and BMX riders must balance are thrust and drag. Thrust is motion in a given direction, while drag—also known as friction—is the force that opposes thrust. Shown: Rayce Davis.

3 Making Science Relevant in the Middle Grades

Students should know what it feels like to be completely absorbed in a problem. They seldom experience this feeling in school.

—Jerome Bruner, *The Process of Education*

Often, students will ask their teacher, "What is the point of this?" They want to know the purpose of the lesson at hand and, in a larger sense, exactly how what they are learning will apply to their real lives. Often, a disconnect exists between experiences in the classroom and their relevance and purpose in a practical and pragmatic sense. This is important in all levels of education, and never more so than in middle school, where motivation to learn—along with content acquisition and conceptual understanding—is a principle teaching goal to help develop an aptitude for student success in education.

For example, physical science concepts at the middle school level are often taught quite traditionally and in an almost clinical manner, with lessons isolated to a specific circumstance in the classroom. Whether using a pendulum or a spring to measure force, often the tools and the content are disconnected from the students' experiences. There is a real need to explore connections of content in settings that are both authentic and relatable for students.

So how is this achieved? In some ways, the teacher has to know the students in the classroom, their interests, and how to integrate those interests into daily lessons. The job of the facilitator is to help learners make connections, and methods for doing this often include the use of analogy, metaphor, and storytelling. In the area of science, these methods are vital; additionally, it is important to integrate students' experiences into the classroom curriculum. The teacher can initiate this by asking the class questions about their own interests; the students' responses can serve as entry points for the teacher to integrate the group's interests into curriculum content. For example, if a student has a keen interest in playing guitar, throwing a Frisbee, or swimming, there are real ways to present scientific ideas that apply to that activity. In swimming, for example, you pull the water back to move your body forward, a great example of a central concept from Newton's Third Law of Motion—for every action, there is an equal and opposite reaction. That is the purpose of action science: to put physical science concepts in the realm of youth culture and, in effect, to make science approachable, relatable, and possibly even cool.

As such, the main content areas in action science—forces, motion, Newton's Laws of Motion, and simple machines—are naturally relevant to students' personal lives, and these connections should be emphasized in the physical science classroom. In each of these overarching focus areas, 8 to 10 specific science concepts are usually linked to the Next Generation Science Standards (NGSS) and aligned with individual state standards, and a teacher can use these concepts to help students explore, engage, and experiment in classroom activities, combined with video instruction. For example, in the areas of forces and motion as applied to middle school science, the central ideas of gravity, mass, balanced forces, unbalanced forces, and frame of reference interact and crosscut in various activities and are strengthened by providing students with hands-on experiences.

In action science, the connection to the content comes from the link to activities students do, see, or know, and can relate to in a fundamental and experiential manner. The overarching theme in this curriculum is to integrate the appeal of skateboarding and other action sports as teaching and learning vehicles for students and science teachers. The main purpose is to provide an interesting method of engaging students in the exploration of science in a real-world context.

Each learner understands content and concepts differently based on his or her previous experiences, and the materials help provide a context for understanding both science concepts and real-world connections. Learners everywhere have an endless array of fascinating content at their fingertips, and with computer access and technology becoming more affordable, even more

When you hear of tricks such as a 360, 540, 720, or even 900, you can now understand that the total number of degrees divided by 360 will give you the number of revolutions accomplished by the rider. Shown: Daniel Dhers.

information is becoming available. The main emphasis here is to engage students in the exploration of science in a real-world context and to link physics to action sports. The students need opportunities to address their own misconceptions and to develop ways to view the concepts as they occur in real-world situations. The teacher has to continue to act as a facilitator in the classroom and therein guide the students toward discovery by using questions to constantly put the responsibility for learning back in their hands. Each student possesses some background understanding of almost any topic he or she comes across in class. Many times, these ideas are correct; other times, they are misguided (Rutherford & Ahlgren, 1990).

Teachers know it is important to engage students in classroom activities that build on their previous experiences and use recognizable materials; students can better understand specific concepts through hands-on approaches (Eisenkraft, 2003). For example, students who enjoy skateboarding can be given opportunities to explore the concepts of velocity, acceleration, center of gravity, and moment of inertia. They may also use the skateboard and a local skate park to investigate topics such as inclined planes, levers, fulcrums, and screws. The purpose of this approach is to allow students to explore meaningful science topics in the context of an activity they find relevant and relatable. As such, the teacher may also integrate other activities of interest and emphasize how the concepts under study can be identified, explored, and calculated in those contexts as well. For example, a drum kit also uses various levers and fulcrums, screws, and inclined planes. The key is to empower students to find the connections in their own experiences and defend them in classroom assignments.

EXPLORING THE CENTER OF GRAVITY IN THE CLASSROOM

As mentioned in Chapter 2, the video segments can provide solid hooks to gain students' interest in the topics as well as an introduction to the content to be explored in the classroom. Prior to showing a video segment, the teacher can pose open-ended questions to activate students' previous knowledge concerning this content. Sample questions in this case might include "What do you do when you ride a skateboard or a bike?" "How do you balance on a skateboard or bike?" and "What forces are acting on you as you are trying to ride a bike or skateboard?" Additionally, previously marginalized students who have experience in these activities but may struggle in science can become experts in this discussion and contribute greatly to the classroom investigations. This effectively activates the engagement phase—the first of the 5Es of constructivism—and puts forward the idea that learning is fun and the student is in control of the learning process.

Finally, the teacher should conclude the series of questions by asking, "What is the center of gravity, and why is it important?" and then use the ensuing conversation to introduce the video that covers gravity, lift, and the center of gravity, accessed with the QR code below.

This short video segment then serves as the engagement for the activity "Flatland BMX and the Center of Gravity" (see Activity 1 in Chapter 8 of this book), in which students create irregular cardboard shapes and determine their center of gravity through a series of step-by-step procedures. Students exploring a concept should be given opportunities to work with hands-on materials so they can have experiences that are real and fundamental. Hands-on learning plays a valuable role in the constructivist paradigm, as it is the process of learning by doing (Dewey, 1902) that is utilized in explorations and experiments.

Next, students modify their shapes by either adding paper clips (which increases mass) or trimming the cardboard (which decreases mass). In turn, they come to see that there is a fundamental relationship between an object's center of gravity and its mass, and that the center of gravity will move in relation to an increase or decrease in mass. After the classroom lesson, the teacher can revisit the activity by asking students to explain their findings and the

Objects in motion and objects at rest both have forces acting on them. Those forces can be balanced or unbalanced. Shown: William H. Robertson.

relationships they discovered. In a constructivist sense, the teacher is activating the exploration phase and encouraging students to venture deeper into the topic and link it to their classroom activities.

With this practical experience, students can begin to articulate the reasons behind their understandings and can compare these understandings within the classroom to the teacher's perspective, as well as to the other students'. This is how the idea of misconceptions in science can be addressed. As in reality, a person will not change his or her position on a topic, whether it is right or wrong, until that person has an experience or set of experiences that verifies this understanding or in effect renders it incorrect (Bybee et al., 2006). As stated in the NGSS, asking questions and defining problems are central skills for students to gain. The teacher's use of open-ended questions provides a framework for this exploration that can be strengthened with a series of investigations to explore concepts in a deeper and self-directed manner. This entire activity can be completed in the time frame of a normal class period, with minimal setup and cleanup, and can provide both the teacher and students with an interesting alternative to exploring these fundamental physics concepts.

INQUIRY ACTIVITIES USING THE IDEAS OF LEVERS AND FULCRUMS

Real-life situations set in a classroom curriculum are fundamental to a constructivist approach in both method and content, and the use of relevant scenarios helps contextualize the concepts, as well as helping provide connections across subject areas (Bentley, 1995; Hofstein & Yager, 1982). Research scientists traverse the barriers between academic disciplines all the time and seldom operate solely in isolated areas of content, instead integrating the use of language, knowledge, and process application. The importance of modeling science practices, identifying core ideas in science, and making connections to crosscut themes and concepts is at the core of the NGSS, and science programs that emphasize investigation give students the ability to retain facts through critical thinking by working through problems logically and making connections to the real world.

The materials in *Action Science* are also designed to emphasize inquiry in classroom explorations. As a foundation for discovery, the teacher can use the video segment that explores the use of levers and fulcrums in skateboarding, and then have the students perform the classroom activity "Skateboards Have Levers and Fulcrums" (see Activity 7 in Chapter 8 of this book). In addition, look at the action science video titled "Dr. Skateboard's Action Science—Simple Machines—Episode 2—Lever & Fulcrum, Wheel & Axle," accessed with the QR Code below.

After the activity, the teacher may revisit these ideas and then create an extension inquiry exercise for the students to complete in teams. The teacher can provide the students with the same materials used in the activity, such as rulers, tape, plastic spoons, rubber bands, and modeling clay, and challenge them to design a simple machine made of at least three of the provided materials that uses a lever and a fulcrum to propel a small marshmallow some distance.

In making this transition in class, the teacher guides the students toward developing their own ideas and, within a given time period, has the students create and test their unique designs. Engaging students in a design competition (to see whose machine can propel a marshmallow the farthest) promotes a spirit of enthusiasm and excitement among the groups; it also provides excellent opportunities for students to develop cooperative group skills and use critical thinking to solve the problem presented. Immersion in a problem is one way to engage fully, to explore, and also to feel a real need to find an explanation. It is also important to develop a sense of self-motivation and to provide a path for self-discovery and, ultimately, content understanding (Bruner, 1962). Finally, the student teams not only have to launch the marshmallow, but they also have to record the distances, calculate the average distance traveled, and identify the lever and fulcrum within their machines. In this manner, the students have to present their ideas, justify their understandings, and support their findings with experimental data.

EXTENSIONS FOR RELEVANCE IN ACTION SCIENCE

As an aside, many aspects of such enhancing learning approaches for middle school students build on the good work teachers regularly do, such as bringing in current events from the newspaper, television, or other popular media sources. Good teachers look to find ways to make each classroom lesson interactive and student centered, such as in leading classroom discussions by asking questions and highlighting content through in-class demonstrations. Being a good teacher means being opportunistic and learning to develop more than confidence in a lesson plan, but also being able to make on-the-fly adjustments, such as when all the class's background knowledge is not quite what you thought it would be.

The inclined plane or ramp is an example of a simple machine that both BMX riders and skaters use. Shown: Art Thomason.

Another example of an innovative teaching approach is the use of large-scale live demonstrations that engage students in multimedia-enhanced stadium settings. Such events have produced video content that can also be used for in-class instruction and motivation in science, technology, engineering, and mathematics (STEM) topics. With the help of a team of professional athletes specializing in the areas of skateboarding and BMX, action-filled scenes that demonstrate specific STEM content have been translated into short STEM-related videos on concepts such as creativity and imagination, integrated with content such as the center of gravity and simple machines. The various members of this "edutainment" effort, with their extensive experience putting on performances in educational settings, form the backbone of each video and serve as a link to the science content being taught.

As the athletes perform each extremely difficult trick, they demonstrate specific science concepts in action, such as the relationships between velocity and acceleration. The use of real athletes provides another pathway to invite learners to learn, in that they may not be initially attracted to science but still recognize and respect the difficulty of the maneuvers performed in the live demo and then captured on video. The videos produced from the demonstrations provide participating teachers and students with a series of instructional opportunities and relevant content information that can be used to explore and explain the given topic, as well as to engage the students in classroom activities.

SUMMARY

The use of educational materials contextualized in the form of action science and grounded in a constructivist methodology has been gaining momentum as content delivery methods have grown to integrate active learning, relevant curriculum, and the use of multimedia and video content. The connection of science content and real-world topics is vital to engage students effectively and to provide them with a reason to delve into deeper conceptual understandings. *Action Science*, with its synthesis of elements of education and entertainment, has great potential to serve as a primary motivational and engagement strategy for academic efforts at the middle school level, especially in physical science education. Additionally, the potential to reach wider audiences in science using these strategies can help transform physical science education by integrating both informal and formal learning in ways that increase student interest and inspire learners to pursue science in school and, ultimately, as professionals.

REFERENCES

Bentley, M. L. (1995). Carpe diem. *Science Activities, 32*(3), 23–30.

Bruner, J. (1962). *The process of education.* Cambridge, MA: Harvard University Press.

Bybee, R. W., Taylor, J. A., Gardner, A., Van Scotter, P., Powell, J. C., Westbrook, A., & Landes, N. (2006). *The BSCS 5E instructional model: Origins, effectiveness, and applications; executive summary.* Retrieved from http://www.bscs.org/sites/default/files/_legacy/BSCS_5E_Instructional_Model-Executive_Summary_0.pdf

Dewey, J. (1902). *The child and the curriculum.* Chicago, IL: Chicago University Press.

Eisenkraft, A. (2003). Expanding the 5E model. *The Science Teacher, 70*(6), 57–59.

Hofstein, A., & Yager, R. (1982). Societal issues as organizers for science education in the '80s. *School Science and Mathematics, 82*(7), 539–547.

Rutherford, F. J., & Ahlgren, A. (1990). *Science for all Americans.* New York, NY: Oxford University Press.

Momentum can be a positive or a negative for skateboarders and BMX riders. It's a positive when a rider hits the ramp at a constant velocity and uses the momentum to perform an incredible trick. Shown: Daniel Dhers.

4 Linking Pedagogy and Science Content in Practice

It is one thing to diagnose the problem, my dear Watson. It is quite another to resolve it.

—*Sherlock Holmes*

As a teacher for middle school students, you might often wonder if it is enough to create situations for learning in science that allow students to engage with and explore concepts without first mastering content. For many of us, content is king, and without an understanding of the vocabulary and basic premises found in science, there is not a clear foundation on which to construct knowledge. In many ways, this is what Bloom was putting forth with his taxonomy: that content knowledge forms the base for learning and that critical thinking skills, such as analysis, synthesis, and evaluation, are logical goals to achieve through a progression of learning from knowledge, understanding, and application.

Yet, for many of us, this is not how learning takes place, as students are not merely content receptors, waiting to receive information openly and without prejudice. In fact, students are in effect problem solvers, looking for opportunities to match what they are learning to what they have learned. This should not be limited to a specific location, such as the classroom, as what happens inside and outside the classroom—in the home, on the streets, with friends and family—is what truly informs learning. Through these experiences, learning takes place, because the things we experience provide a foothold for making sense of the content.

When I was in high school and had just received my driver's license, my father wanted me to be prepared for the unexpected when driving the family car and felt it was important that I understand how to change a tire in case I had a blowout or got a flat. My father took me aside one day and offered this instruction in our driveway, showing me where to find the tools and spare tire in the trunk.

He went over the basics of safety, explaining how to put something under the wheels to ensure that the car would not roll, showing me where on the car frame to insert the jack, and instructing me to place the jack level on the ground. He showed me in a hands-on manner how to take off the tire and put on the spare, using memorable phrases to help me remember which way to turn the lug nuts with the tire iron: lefty-loosey and righty-tighty, meaning I should turn the nuts left to loosen them and right to tighten them.

In due time, my father had guided me through the process from start to finish. I had replaced the tire with the spare and then repeated the steps to return all the tools and the spare tire back to their place in the trunk. I had completed a dry run for changing a tire, a rehearsal under the guidance of a teacher who carefully provided feedback in a safe environment.

I had performed this exercise in the safety of my own driveway, and exactly when do you think I got my first flat tire? Do you think I was in my driveway? Of course not. The first time I got a flat tire, I was driving down the freeway outside of the city. The blowout forced me to move quickly to the side of the road. With the car stopped and other motorists whizzing by, I had to recall the lessons I had learned from my father in the safety of our driveway. In many ways, I was unprepared for the open-ended nature of the task at hand, although I had some direct content knowledge and guided practice.

I struggled with the car, the location of the tire, the tools, the jack, where to place the jack—all the while trying to be safe on the side of the road. In due time, I was able to change the tire, and I drove the car straight home and enlisted the help of my parents to see if I had done it correctly. Needless to say, I was exhausted and had been through a situation of high risk and high ambiguity. In one sense, that was the day I truly learned how to change a tire.

For our students to experience real learning, we have to be willing to take them to areas of high risk and high ambiguity within the safe environment of our classrooms and provide them with authentic tasks in context that require them to leave their comfort zones. I am not advocating that we increase stress or make our classrooms unsafe; I am promoting the idea that we need to stretch the way students learn and integrate more open-ended approaches that emphasize problem-solving strategies in class.

The modern skateboard turns up at the nose (or front) and at the tail (or back), which forms two levers. The levers on a skateboard allow the rider to perform G-turns, nollies, tailslides, and blunts, to name just a few choice tricks. Shown: William H. Robertson.

LINKING CONSTRUCTIVISM TO THE INTERNET

Getting students engaged and exploring concepts is often not enough, especially when they are expected to master content, not merely be exposed to it or work with it in purely prescriptive approaches. It is important for teachers to design authentic tasks that allow students to gather and analyze data to uncover content relevant to the ideas they have been exploring (Eisenkraft, 2003). Often, classroom teachers face the pressures of high-stakes testing and covering massive amounts of material in a limited time. Many students are disengaged and do not seem to enjoy learning the skills, nor do they see connections between the real world and the topics being studied. By implementing the constructivist approach of the 5E lesson model in an innovative and creative way, students are engaged in learning the required science skills and participate in higher cognitive interactions (Ergin, Kanli, & Unsal, 2008). Just as scientists constantly communicate with one another to solve problems, students should be engaged in the higher-order thinking skills that include synthesis, evaluation, and application of information—not memorization (Shepard, 1989). Real-world situations are unique, and solving them often requires new methods or techniques. Problem-solving strategies often change along with the underlying concepts (Bruner, 1962).

As part of the explorations students conduct in class, they should be encouraged to investigate science topics and related areas of interest on the

World Wide Web. The Internet can also be a source for content that explains and elaborates on areas of constructivism. The Internet should be a comprehensive part of the research and dissemination of subsequent student products. Ultimately, in the evaluation phase, students should also publicly demonstrate their understandings, and posting their work on the Internet is one strategy that can be quite useful. This links well to required learning outcomes in knowledge and skills, as informal evaluations have shown that students who publish their work on the Internet increase their reading and writing skills (Herman, Osmundson, & Pascal, 1996). One reason for this may be that work on the Internet is available for the world to see at any time; thus, it stands to reason that a student who knows this will prepare his or her work more carefully and, in so doing, increase writing and reading abilities.

The wheels and axle on a BMX bike can also serve as fulcrum points on the bike's frame to enhance maneuvers. Shown: Billy Gawrych.

Ultimately, students should be engaging and participating both inside and outside of class, as this is crucial to learning and the construction of purposes and meanings (Wiggins, 1989). The teacher should actively promote and encourage positive group interactions and cooperative behaviors that foster the types of thinking interactions that enhance the learning process.

STRATEGIES TO ENHANCE STUDENT PROBLEM SOLVING WITH TECHNOLOGY

In an effective classroom, learning requires more than connecting new material to old ways of thinking; it also must connect new material to new ways of understanding. Communication from and with multiple peoples and perspectives is important and vital to learning (Orgill & Thomas, 2009). Students need experiences that help them develop new views and make better sense of their world (Bybee et al., 2006). In the classroom, the teacher can facilitate online discussion, invite local community members to act as electronic mentors, and even, in this day of great connectedness, invite experts in specific fields, such as physics, to answer questions posed by students in class.

Within this approach, each student should be tasked with designing, developing, and reflecting on a current real-world topic that integrates specific science concepts. In describing and explaining ideas to others, the learner synthesizes material in a way that requires higher-order thinking. A person who can successfully explain a body of knowledge to others may be said to have mastered that body of knowledge. The purpose is to have the learner demonstrate an understanding of problem-based learning as associated with inquiry-based science facilitated in a web environment.

Student products that integrate technology can have great value in a classroom, and if the teacher offers students choices, the results can be quite impressive. An example of this would be having students develop a mind map or concept map that shows how the content they are working with fits within their understanding of a specific concept related to motion, such as a pitcher releasing the ball to a catcher, a dancer balancing artfully on one leg, or a basketball player making a long shot with an enhanced trajectory arc. Students would deconstruct each of these activities—and students could easily come up with more of their own examples—in terms of various science concepts found in the image and how those concepts connect and interrelate. Having the students share this work, rehearse their responses in small groups, and defend their positions based on their conceptual understanding aligns clearly with the science and engineering practices found in the Next Generation Science Standards (NGSS).

To have success in learning science with an open-ended framework, which relates to emphasizing the engaging and exploring aspects of constructivism within the context of a problem to solve, the learning environment should promote the use of telecommunications for research and collaboration with other participants. For example, students can demonstrate their acquired knowledge by accessing functional and well-designed websites prescreened by the teacher and made available as bookmarks on classroom computers. This helps bring the issue alive and engages the students in active learning. Appropriate software and training are essential in providing a positive telecommunications environment for all students, while sound technical support is vital in keeping the lines of communication open.

Web-based technologies in this example can be defined as computers with Internet access, multimedia tools, video, and software programs that will aid in the acquisition, collaboration, and communication of information. The use of technology to foster and develop critical thinking skills is an area of great potential in education. Researchers have indicated that technology as it has been defined in this book, when integrated properly into a web-based curriculum, may very well improve students' problem-solving skills, as well as their ability to apply critical thinking to the information they access on the Internet.

Web-based learning activities may include completing a webquest, in which students investigate a topic and answer specific questions using teacher-defined websites, or finding a website that integrates science and the activity they are investigating (the connections on the web are as wide open as one's imagination) and then sharing the material and evaluating its accuracy based on a content comparison to what they have been learning under the teacher's guidance in the classroom. Such tasks further strengthen the core ideas in the discipline and help the students further connect concepts not only in the classroom but also across the fabric of their own life experiences.

Technology can be the modern teacher's ally and should be effectively integrated into the presentation and demonstration of the curriculum. This takes a different style of teacher, one who learns from students and also models the use of technology in the classroom (Duffy, Roehler, Meloth, & Vavrus, 1986). Today's student needs to be stimulated, and since technology is an integral feature of the modern world, failing to use it in the classroom is a real disservice to students. In science, "technology provides the eyes and ears of science—and some of the muscle too" (Rutherford & Ahlgren, 1990, p. 26). Technology, whether a computer or a calculator, is vital to teaching the concepts associated with data collection, computation, and measurement—a fact that is reinforced in the NGSS and Common Core State Standards Initiative in Mathematics.

This is the point where the Internet is most powerful and the motivation to do good work becomes intrinsic and not driven by the pursuit of a grade. It is one thing to complete a project and turn it in to a teacher in your school; it is quite another to publish your work on the Internet for anyone with web access to read and consult.

The textbook is a classroom resource but certainly not the only resource, and the nature of knowledge should incorporate multiple viewpoints and sources that include textbooks, the Internet, multimedia, and other outlets for current information. Knowledge is as much about process as it is about content, and the two must be integrated effectively so the learner sees the value of the content in a conceptually correct context. Students should explore multiple examples from many cultures and time periods and should be given the time to make sense of it all. The goal is to engage the learner in higher-order thinking that includes analysis, synthesis, and evaluation of material and information.

For the student, the content included within links to websites should consist of primary content sources such as research papers, associated activities that fit into the scope and sequence of their projects, and appropriate reference materials. Students develop a background for learning that matches their own interests using the appropriate technology, share their information with others in the classroom, and present this information in class for assessment. The purpose is to have the student demonstrate an understanding of the science concepts associated with a unit of study and to prepare the teacher to provide students with the opportunities to discover content that links to their own experiences and areas of interest. Building a foundation of problem solving, not one of content memorization, within a real-world science topic requires that each student include appropriate research and supporting activities within the content of their open-ended explorations that will provide the background for activities to follow.

In skateboarding, the ollie is a trick that embodies the essence of flight, in that a rider must lift himself or herself into the air and travel forward with a bit of momentum. Shown: Rayce Davis.

SUMMARY

In our classrooms today, there is great emphasis on achievement in the areas of science, technology, engineering, and mathematics, and at the middle school level, technology can be a great tool to integrate within a teaching framework centered on constructivist principles. Content available on the Internet, including videos and textual materials, is quickly becoming the primary source for information, and integrating this into the science classroom is critical. An important role of the teacher is to allow students the time to explore, research, and discover content on the web, and to verify its accuracy through experiences in the classroom. Science is often seen as an area where ideas are born and should be viewed not as a place to find the right answers but as the process by which one works to eliminate what is false and thereby move closer to what is true. Technology is often called the muscle of science, as it can allow students to access information as well as explore concepts in ways that are both imaginative and informative. In action science, the video content is an inherent part of combining technology with instruction in a constructivist framework, and it also provides the teacher with the room to be creative in using technology to encourage students to dive deeper into topics of interest. Ultimately, technology is part of our lives, and no more so than with those of the digital-native generation currently in middle school classrooms everywhere. The use of technology should be seen as an integrative part of the curriculum.

REFERENCES

Bruner, J. (1962). *The process of education.* Cambridge, MA: Harvard University Press.

Bybee, R. W., Taylor, J. A., Gardner, A., Van Scotter, P., Powell, J. C., Westbrook, A., & Landes, N. (2006). *The BSCS 5E instructional model: Origins, effectiveness, and applications; executive summary.* Retrieved from http://www.bscs.org/sites/default/files/_legacy/BSCS_5E_Instructional_Model-Executive_Summary_0.pdf

Duffy, G., Roehler, L., Meloth, M., & Vavrus, L. (1986). Conceptualizing instructional explanation. *Teaching and Teacher Education, 2*(1), 1–18.

Eisenkraft, A. (2003). Expanding the 5E model. *The Science Teacher, 70*(6), 57–59.

Ergin, I., Kanli, U., & Unsal, Y. (2008). An example for the effect of 5E model on the academic success and attitude levels of students': "Inclined projectile motion." *Journal of Turkish Science Education, 5*(3), 47–59.

Herman, J., Osmundson, E., & Pascal, J. (1996, October). *Los Alamos National Laboratory critical issues forum final evaluation report.* Los Angeles: Center for the Study of Evaluation, UCLA Graduate School of Education.

Orgill, M., & Thomas, M. (2009). Analogies and the 5E model. *The Science Teacher, 74*(1), 40–45.

Rutherford, F. J., & Ahlgren, A. (1990). *Science for all Americans.* New York, NY: Oxford University Press.

Shepard, L. A. (1989, April). Why we need better assessments. *Educational Leadership, 46,* 4–9.

Wiggins, G. (1989, April). Teaching to the (authentic) test. *Educational Leadership, 46,* 41–47.

Riders on ramps or at the park move between moments of kinetic and potential energy. For example, while rolling up the ramp, the rider demonstrates kinetic energy, and while stalling at the top, the rider shows the presence of potential energy. Shown: Rayce Davis.

5 Using the 5Es in Action Science

Though intrinsic value is no criterion for a puzzle, the assured existence of a solution is.

—Thomas Kuhn (1970)

Physics is important in many areas of our lives, and we use the principles of forces and motion as we go about our daily activities. For students, the connections to physical science concepts, such as center of mass, acceleration, simple machines, and motion, need to be defined in practical and relatable ways. In the classroom, the teacher can employ the constructivist methods and use a 5E framework while implementing activities of interest that can help increase motivation to learn, along with making science cool in the classroom.

As stated in Chapter 2, constructivism can be characterized in a five-phased process known as the 5Es, which includes the phases of engagement, exploration, explanation, elaboration, and evaluation. As the strategies and materials are implemented with students, the use of a constructivist strategy set in the context of youth-centered activities, such as skateboarding and BMX, proves to be of great interest and impact for the students. The 5E approach as a way of organizing a curriculum has been demonstrated to be effective in presenting concepts in understandable and relatable increments that give students structure and provide integration within content and instruction (Brooks & Brooks, 1993). What follows is a synopsis of how the 5Es of Constructivism and the principles of action science are integrated into an educational, student-centered learning experience.

As the events of action science have been lived out with students, the use of a constructivist strategy set in the context of a youth-centered activity (e.g., skateboarding) proved to be of great interest and impact for students. Not only were students engaged, but they also participated actively in deepening their own understandings of concepts in physical science. The 5E approach should be seen as a way of organizing a curriculum that can provide students and teachers with a method for teaching and learning with purpose (Eisenkraft, 2003).

For example, a skateboard or bike can be a great platform to introduce and solidify the concepts of simple machines and the differences between simple machines and complex machines. Simple machines include any device that makes work easier with a single motion. Simple machines do not allow you to do less work, but they do make the work easier in three main ways: (1) by lessening the force exerted, (2) by changing the distance over which a force is exerted, or (3) by changing the direction of the force exerted. Although not all simple machines are used in skateboarding and BMX, a number are present.

BMX riders need to manage specific factors, such as force, centrifugal force, and moment of inertia, to land cool tricks. Shown: Daniel Dhers.

A teacher can begin a lesson by holding up a skateboard in class. This serves as a hook for gaining student interest, as well as presenting an opportunity to engage students in the content of simple machines. The teacher can begin by asking students to identify the three simple machines on a skateboard: the lever and fulcrum, the screw, and the wheel and axle.

The teacher can then walk students through the elements of these three simple machines and continue to ask them where these machines can be found. The modern skateboard turns up at the nose (or front) and at the tail (or back). The upturned kick acts as a lever for the rider and helps lessen the force the rider exerts while performing tricks on ramps, in the street, or on level ground. The place where the trucks of the skateboard and the deck come together is an example of a fulcrum, or a fixed point around which a rigid lever moves. The fulcrum action allows the rider to control the movement of a trick by applying or releasing pressure on the fulcrum point.

Another simple machine to explore is the wheel and axle. On a skateboard, this refers to the urethane wheels complete with sealed bearings and the trucks. On a skateboard, the wheel-and-axle simple machine allows the rider to roll, carve, grind, and spin. Wheels and axles also make changing the direction of motion easier so that forward or backward motion can happen quickly. The wheel and axle can be found in a number of modern technologies, including motorcycles, cars, buses, and airplanes.

The screw is another simple machine found in both skateboarding and BMX. On a skateboard, the screws hold the deck to the trucks. A screw can be defined as an inclined plane wrapped around a cylinder. The screw allows a skateboarder to use the levers while rolling along on the wheels and axles. The screw action also keeps the wheels on the axles and helps the rider control the turning radius of the trucks.

ENGAGEMENT IN ACTION SCIENCE ACTIVITIES

In this book, eight sample activities cover the areas of forces, motion, Newton's Laws of Motion, and simple machines. If we look at the open-ended discussion on simple machines and the teacher's introduction of specific concepts using the skateboard as a form of engaging students, we can see how important it is to increase motivation for learning in the classroom and that the connections to previous experiences are vital to activating conceptual understandings.

From that point, the teacher may want to use the classroom activity "Around the World With Wheels and Axles" (see Activity 8 in Chapter 8) and present the associated video clip on simple machines. The video clip thus becomes a vehicle for further engagement, helping define for students the coming classroom task. The teacher should follow the video clip with questions such as "What concepts about simple machines were shown in the video?" "Where is the wheel and axle on a skateboard? On a bike?" and "What other objects that you encounter in your life have wheels and axles?"

The video content is presented to students not only to demonstrate complex athletic moves using the skateboard but also to introduce the concepts associated with wheels and axles, which the students will explore in greater depth in the classroom activity. The engagement aspect of constructivism should help the students make connections between their past learning experiences and those to be had in the classroom. Students should be able to see clearly exactly which concepts they are to explore and how they will investigate them in the classroom, which, in turn, leads to the next step in the constructivist framework. For critical thinking to guide learning, the students must initiate a plan and take action, and this process in constructivism begins with engagement.

EXPLORATIONS OF SCIENCE CONCEPTS

A quality exploration activity is important and central to building on the initial aspects of getting students engaged in a given topic of study. In the case of the students engaged in the wheel-and-axle activity, a defined experimental procedure to be done in small groups incorporates the need for teamwork and requires gathering data. The students should be divided into teams of about four members, with the teacher serving in the role of facilitator and also modeling the procedure to be followed.

In a constructivist framework in our example activity, the exploration phase should give students direct experiences with wheels and axles and allow them the freedom to discover how wheels and axles really work. The materials in the activities are familiar, and the procedure requires a hands-on, minds-on approach.

Exploration experiences in this manner rely on establishing real-world connections, using materials and technology for hands-on interaction, and providing a common base of experiences from which to grow and learn (Bybee, 2003).

Momentum can be a positive or a negative for skateboarders. It's a positive when a rider hits the ramp at a constant velocity and uses the momentum to perform an incredible trick. Shown: Lewis Dinsdale.

The simple machines formed by the wheels and axles also directly connect to concepts found in force, which can be calculated in the "explain" section of the activity.

The teams' task is to gather data from two different trials and compare the forces on an object with wheels and axles to forces on an object without wheels and axles. Each group should do each trial three times to give students a data set from which they can draw conclusions based on calculations and overall averages. As a result of this approach within the exploration aspect of the wheel-and-axle activity, the fundamental task of the student teams is to collect and analyze the data in two separate trials. This requires the teams to work together cooperatively, to be in charge of their own learning, and to search the content of the activity to make connections between what they are measuring and the science concepts they are asked to utilize.

EXPLANATIONS OF TOPICS LINKED WITH DATA

In a constructivist framework, the explanation phase helps students uncover the content surrounding the concepts they have been exploring. Students should now have opportunities to verbalize their conceptual understanding, to encounter new content material, or to demonstrate new skills. This phase also provides opportunities for teachers to introduce scientific and mathematical primary content materials such as formal terms, definitions, and other content information. "Many scientific explanations are not intuitive, particularly to students who are encountering concepts for the first time" (Orgill & Thomas, 2007, p. 42). The implementation of this phase provides the learner with opportunities to identify skills and behavior both to experience and to discover content that may be useful in context.

The explanation phase should also allow students to develop skills and behaviors that will help them be successful in their learning. Students, like revolutionary scientists, need experiences that help them develop new views and make better sense of their world. Learning is the responsibility of the learner, but the teacher guides the learner into developing meaning from content material and classroom experience. Communication from and between multiple peoples and perspectives is important and vital to learning.

In the activity, student teams are asked to work together on the data collection and to answer the associated questions found in the "evaluate" section. At the end of the class period, the student teams should be able to complete their calculations and solutions and provide them to the teacher, who in turn can use this information as a formative assessment to provide feedback (Stephens, Bottge, & Rueda, 2009). The teacher can use this opportunity to facilitate, modify, and differentiate elaboration or additional action science activities later in the week to meet students' individual needs. It also gives the teacher insight into what concepts the students understood well and which might require more time and enrichment opportunities to be fully understood.

In this sense, the student teams have to interact with one another and the content of the activities in potentially unanticipated and unpredicted ways, and they must negotiate connections to the content not only individually but also as a group. A person who successfully explains a body of knowledge to others may be said to have mastered that body of knowledge. In describing and explaining ideas to others, the learner synthesizes material in a way that requires higher-order thinking (Ergin, Kanli, & Unsal, 2008). The key to this in terms of the wheel-and-axle activity is that students have an experience first and are then asked to manipulate the content, rather than first being provided with content information and then being asked to develop an experience or complete an activity with that content.

ELABORATING ON CONCEPTS BY EXTENDING ACTIVITIES

Following this activity, the teacher has opportunities to dive deeper into a topic as well as provide options for learners to extend their own learning. In a constructivist framework, the educator provides chances for learners to practice and refine their skills and behaviors in authentic contexts. Students are also given multiple opportunities to deepen and broaden their knowledge base and integrate that knowledge into their conceptual understandings and actions, both inside and outside of the classroom (Martin, 2009). This instructional strategy allows the student to spend time exploring and explaining the process, with time for reflection and numerous opportunities to synthesize information.

Through new experiences, learners develop deeper and broader understandings of major concepts, obtain more information about areas of interest, and refine their skills. In classroom settings, constructivist educators in the elaboration phase introduce variables that students can explain in deeper ways. Issues can be examined from multiple perspectives and cultural viewpoints. The key is to build on previous learning and apply new learning in a meaningful context. Mastering content does not imply well-constructed knowledge, but the application of content within conceptually correct frameworks does, ultimately promoting understandings with impact.

During the work in the classroom, the students can be challenged to deal with more complex physical science topics associated with simple machines and their relationship to forces and motion. For example, in the area of levers and fulcrums, the students might construct a catapult and calculate distances and the angle of release for each trial. Another example of an elaboration extension in the area of forces is to have students construct a hovercraft out of cardboard and other simple materials, and use a balloon to provide a cushion of air for movement along the table. The students in this case might have to calculate the distances traveled and understand the scientific relationships among forces that are related to solving these problems, and describe processes using graphs, tables, and models.

In the wheel-and-axle example, the teacher can display pictures of Stonehenge and hypothesize how such large stones could have been moved and positioned in such a way. The teacher can then ask students to use what they know about simple machines to determine how these huge slabs of stone were put into place without the use of modern machinery. Students should be reminded that the rollers or logs they used to move their book are similar to the wheel and axle.

Newton's First Law states that objects in motion will remain in motion unless acted on by an unbalanced force and objects at rest will remain at rest unless acted on by an unbalanced force. Shown: Art Thomason.

The teacher can discuss the issues of wheels and axles and ask students to write essays about how the wheel has helped humans. Students can then make a list of the many uses of the wheel or brainstorm as the teacher makes a list on the board, or enters the results of the brainstorming activity into concept-mapping software. Students can then pick a topic from the list to build into a project, and report on exactly how the wheel and axle are used in that area to make work easier.

EVALUATION TOWARD AN UNDERSTANDING OF CONCEPTS

Finally, at the end of the activity, the students should be given not only the opportunity to answer the questions for the activity but also a chance to evaluate their own learning and to write reflections about what they have learned in class that day. This is vital to gauging students' understandings, but it should also be pointed out that both types of student self-evaluations show a balance on both results outcomes and process skills development. As such, students learn to assess their own abilities, identify areas of mastery they now possess, and strengthen developing understandings and abilities.

Additionally, students can peer-review the work of others, share their own work, get feedback from others, and self-assess their work to identify areas that need strengthening. The important point is that the learners search to understand what they know and defend that construction of knowledge so the teacher and experts in the field accept it as conceptually correct.

In the case of action science, the activities presented and completed in a constructivist framework give students a chance to provide feedback on the impact of the activities. After the day of activities, the teacher and the whole class can assemble to debrief and review comments and feedback from students to better understand students' gains and needs for improvement in the content focus areas. In turn, students will then be able to discuss their findings with their colleagues and the teacher and, ultimately, to explain the concepts of science and connect them to skateboarding.

SUMMARY

In this section, the practical aspect of the 5Es of Constructivism can be seen within the activities of action science. Primarily, the phases of constructivism—engage, explore, explain, elaborate, and evaluate—are couched within an activity on wheels and axles as simple machines. For the teacher, the goal is to connect each learning opportunity to the students' world and excite the discovery process by building on experiences that allow students to uncover concepts in a hands-on manner.

Have you ever thought of a student riding a skateboard or BMX bicycle as a physicist or engineer? In many ways, the traits of a scientist are the same as those of BMX riders and skateboarders. A physicist or engineer studies the interactions between concepts in physical science and attempts to create situations where the output forces are maximized by the input forces. Simple and complex machines help scientists, as well as skateboarders and BMX riders, maximize their input forces to make a given task easier. Both skateboards and bicycles are examples of complex machines, which are two or more simple machines combined.

REFERENCES

Brooks, J. G., & Brooks, M. G. (1993). *In search of understanding: The case for constructivist classrooms.* Alexandria, VA: Association for Supervision and Curriculum Development.

Bybee, R. W. (2003). The teaching of science: Content, coherence, and congruence. *Journal of Science Education and Technology, 12*(4), 343–358.

Eisenkraft, A. (2003). Expanding the 5E model. *The Science Teacher, 70*(6), 57–59.

Ergin, I., Kanli, U., & Unsal, Y. (2008). An example for the effect of 5E model on the academic success and attitude levels of students': "Inclined projectile motion." *Journal of Turkish Science Education, 5*(3), 47–59.

Kuhn, T. S. (1970). *The structure of scientific revolutions.* Chicago, IL: University of Chicago Press.

Martin, G. (2009). *Focus in high school mathematics: Reasoning and sense making.* Reston, VA: National Council of Teachers of Mathematics.

Orgill, M., & Thomas, M. (2007). Analogies and the 5E model. *The Science Teacher, 74*(1), 40–45.

Stephens, A., Bottge, B., & Rueda, E. (2009). Ramping up on fractions. *Mathematics Teaching in the Middle School, 114*(9), 520–526.

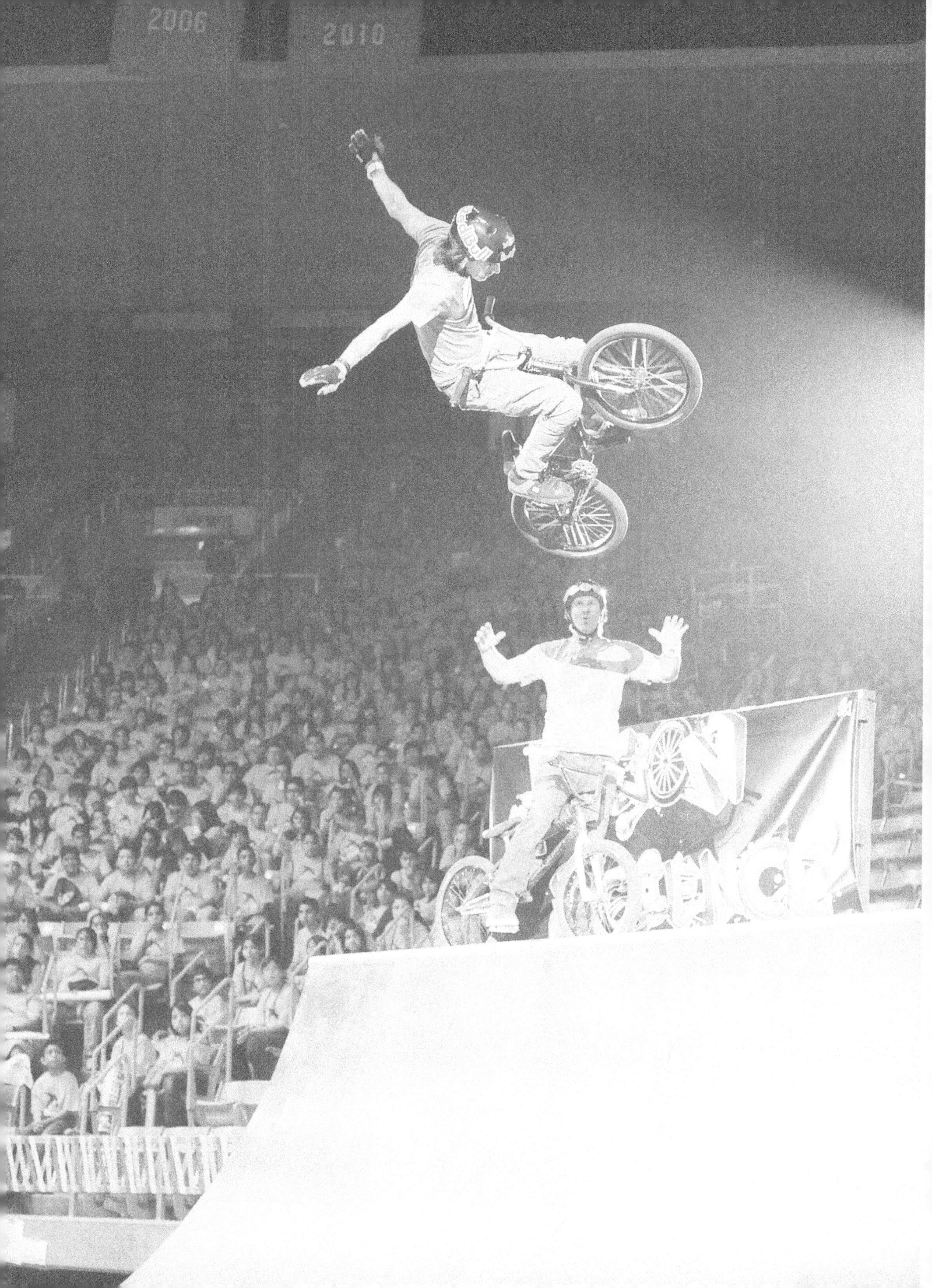
2006
2010

When a rider is in motion on a ramp, the potential energy from dropping in is transferred to kinetic energy at the bottom of the ramp. As the rider travels down the ramp, the potential energy decreases as the kinetic energy increases. Shown: Rayce Davis.

6 Unlocking Resources for Active Learning

Learn. Unlearn. Relearn.

—*Cathy Davidson (2011)*

I have always felt that the best way to learn is to be motivated and find a personal connection to the content based on experience. As a teacher, I have seen that the constructivist methodology provides a framework to emphasize educational opportunities based on doing things first and then looking to explain them once an interest and a reason to learn have been achieved.

In the area of action science, the idea of motivation is central, for both students and educators alike. The goal is to give the middle-grade teacher a set of resources that are conceptually correct in an area generally overlooked and underemphasized in the middle school classroom—that is, physical science.

Much of this, in my experience, is due to the fact that most teachers in Grades 6 through 8 feel less appropriately prepared in this area of science, and most teacher preparation programs do not devote extensive content development in the area of physics for middle-grade educators.

Additionally, as a teacher, my motivation was in part to provide such a resource for the classroom teacher, a resource that would help in practical terms and provide a platform for engaging students in activities that are both meaningful and purposeful. Yet my primary motivation was for the middle school students, specifically those who were somewhat marginalized or disinterested in learning due to the seemingly rote nature of most classroom science curricula. I wanted to see if I could find a way to extend the learning to the students, who in turn would take it into the world in which they live. In that way, education was not limited to the classroom, and the students could see that they were in charge of their own learning, much like the central idea of constructivism: The learner is ultimately responsible for his or her own learning (Dewey, 1902).

Let me relate a story that not only provides the context for this personal motivation for students to learn, which I believe is a common element for all teachers, but also shows that they care and have great concern for their students, feeling understanding and empathy for each learner and individual who enters their classroom.

LIVING THE LIFE IN ACTION SCIENCE

On Saturday morning, as I regularly do, I headed over to the local skate park to get in a session and enjoy a beautiful spring day in the southwestern United States. I enjoy going to the parks on weekend mornings, as it provides me with a chance to get in some good exercise and hone my skateboarding skills. It also provides me with a platform and opportunity to interact with young people and to stay in touch with the pulse of skateboarding. As I go about my morning routine of stretching, checking out my board, and putting on my pads, I also check out who is at the park and where the best lines will be for the day. I choose the mornings because the park is usually less crowded and also has a good mix of families, quality skaters, and, of course, a number of more "mature" skateboarders such as myself.

Anyway, I was enjoying this particular morning, setting into my routine of warming up in easier areas of the park and then moving into the more advanced areas. I noticed a small group of skaters, about 13 years old, seemingly following me as I moved from one area to another. This was not a group I was used to seeing at the park, and the way they were lingering nearby without really interacting with me was a bit odd. As I moved to another area, I noticed that two of the boys were now riding by me more quickly, more closely, and more regularly. I also noticed that they were giving me the "eyeball," kind of checking me out in a way that held an element of confrontation. Now, I know I stand out at the park, as I am a "big guy in a helmet" and well over the median age of the normal skateboarder, but this was somewhat different.

What happened next both inspired and amazed me. One of the young skateboarders rode up to me and screeched his board to a stop. He said, "You're Dr. Skateboard, aren't you?" I responded that, yes, I was Dr. Skateboard and I often came to the park to ride on the weekends. He then turned to me and stated, "I thought so! You came to our school!" At this time, the whole group of skaters had joined him, and I responded, "Really, what school do you attend?" He then mentioned the name of his school, and while I recognized the name and I do a lot of demonstrations at local schools, I realized that I had not been to his school in person. I said, "I don't think I have come to your school to do a demo." He responded quickly, "You didn't come to my school to do a demo. You came in your video!" Then a second skater screeched into the conversation and blurted out, "Yeah, and we're learning about forces and motion!" They said all this with a lot of attitude and were sort of in my face. Their manner caught me off guard a little.

After proper introductions all around, I set out with this small group of skaters, all of whom were in middle school, and we rode around the park and talked about physics concepts. We were no longer just a group of skaters; we were scientists talking about real concepts, such as momentum, acceleration, velocity, and inertia. We looked around the park to identify the simple machines on our boards and the inclined planes all around us. In that magical moment, I had the good fortune to realize my motivation for putting forth the ideas of action science—these methods might actually appeal to young people and inspire them to learn outside of the classroom, making connections to and taking ownership of their own learning.

MAKING THE CONNECTION LAST FOR TEACHERS AND STUDENTS ALIKE

As you can see from the story above, making connections with students is important inside the classroom but especially important outside the classroom. It is also critical for teachers to have resources and methods to reach out and help all students succeed, particularly those who are marginalized or disadvantaged due to limited resources, life situations, or geographic position (Lee, 2002). In the case of action science, one large mission inherent in this approach is to find pathways for learning science that extend into the real-world experiences of all students and also to ignite a love of learning science in an approachable and supportive manner.

Students exploring a concept should be given opportunities to work hands-on with materials so they can have experiences that are real and fundamental. It is important for teachers to engage students in learning situations that effectively integrate their own experiences and familiar materials that students can use to better understand specific concepts (National Research Council, 2006). For example, students who enjoy skateboarding can be given opportunities to explore the concepts of velocity, acceleration, center of gravity, and moment of inertia. They may also use the skateboard and a local skate park to investigate topics such as inclined planes, levers, fulcrums, and screws. The purpose of this approach is to allow students to explore meaningful science topics set in the context of something they enjoy doing.

For the educator using the materials in this book, a series of videos can also be used to engage students and, I hope, to help extend the learning from the textbook into the lives of those in the classroom. For example, in the video on forces that focuses on the concepts of thrust and drag, the action demonstrated shows how riders in skateboarding and BMX effectively use these opposing forces. In the videos, text explanations of the concepts overlay the action, with arrows of different colors placed to show the direction of each resultant force with respect to the tricks the riders are performing. In addition to this visual content, an audio narration reinforces these concepts and helps contextualize the action within the focus of physical science concepts pertaining to forces. The combination of visual, oral, and video content delivers a set of engaging material that comes across as learner centered and relevant.

The premise of this approach is that as students engage in a curriculum that integrates their real-world interests, their understandings of science content will increase. Additionally, the underlying message is that the characteristics that make action sports athletes successful, such as practice, persistence, dedication, and setting goals, also extend to the student's individual academic and educational pursuits. In other words, the fundamental message regarding the importance of education demonstrates that skaters, just like any other students, can be smart and need to set educational goals that fit with their experiences and aspirations, not just their zip code or location in life.

Along with the videos available for free on the web, Chapter 8 of this book also includes two classroom activities for each of the four theme areas (forces, motion, Newton's Laws of Motion, and simple machines). The activities all use common household materials, such as paper clips, card stock, string, and tape. This chapter provides teachers an affordable and practical series of experiments that engage students in hands-on explorations of the physics concepts presented in the video instruction. As the activities were developed, they were field tested in actual classrooms and revised based on interactions with the 15 teachers involved in the design, development, and implementation of the activities with students. The activities are aligned with Next Generation Science Standards for middle-grade physical science in the outlined topic areas.

So what does this mean for the classroom teacher? Isn't this something most teachers are already doing? In many ways, there is nothing new under the sun in education, and most teachers are working to engage students in content that connects to their everyday world. Yet today, with the integration of technology into classrooms, there is also a chance to change the paradigm to better match the ways students learn individually, not just in terms of how the class average appears or in the way a curriculum is laid out in scope and sequence (National Science Board, 2006). Action science is a way for teachers to adopt an attitude of relevance and to seek out content that connects to the students, not merely to recycle the lessons and activities they have used each year without revision and, in many ways, in constant repetition irrespective of students' learning styles or individual strengths and weaknesses.

To this end, the union of action science in the constructivist framework emphasizes the need to provide relevance in the curriculum, and the teacher has to know the interests and motivations of the students before lessons can be

There are three basic types of energy as it relates to motion. First, there is the energy in motion, or kinetic energy. Second, there is the energy of position, or potential energy. Third, there is the total energy of an object, called the mechanical energy, which includes both potential and kinetic energy. Shown: Daniel Dhers.

connected and extended outside the fabric of the classroom (Bailer-Jones, 2002). For example, a student who has an interest in music, especially a modern expression such as rap, can be inspired to use his or her talents to portray content in a different medium. Using concepts in physical science and the technology of video or mixing music on the computer, the student can demonstrate individual understanding with the guidance and support of the teacher. The teacher also has to have enough confidence to allow this to happen in the classroom and to lean on the students' expertise to help supplement the curriculum. No longer needing to be the expert, the teacher manages the flow of the learning, sets the expectations for learning, and allows students to engage actively with one another as they explore the content.

The actual content knowledge students gain in mastering a test has little to do with their demonstrable understanding, yet this is the way most teachers have to respond. The fact that students need to take tests and achieve scores related to their learning is not the point of contention; rather, it is that, in learning, we often emphasize the test content and not the concepts in the curriculum and how they best connect to the students' lives.

To put it in the context of today's middle school learner, let's go back to the example of the students I met in the skate park, described earlier in this chapter. My experience with the learners of today is that they do not learn the same way I learned in school. For me, the content was delivered, I memorized it, and, in effect, I regurgitated it for the exam and received a high grade. I was a functional education parrot, programmed to redeliver the same information in the same way it was provided to me. That won't work with the students now, not when they grow up immersed in a visually enriched environment that in many ways deemphasizes the written word, reducing it to symbols and context—in other words, making it a connection to an experience.

The students at the skate park look first to the realm of video for instruction, using free YouTube videos, for example, as a way to learn how to do most anything, whether it's performing a trick on a skateboard, playing a riff on the guitar, tying a bowtie, or solving a math problem. The use of video as a primary content delivery mechanism is here, but for most folks in education, it is still just a supplement to the content. For most teachers, the written word, the text, is still the place to start. The argument here is that today's students begin with the video to see how things are done (the engagement and exploration) before they ever google the content (the explanation). In other words, the students are demonstrating with technology that they are constructivist in their learning, and teachers have to recognize that fact and embrace it.

This is where incorporating the students' interests and allowing them to engage with and explore the content using technology can bring huge gains in learning, as the teacher can focus the students on concepts in class that are found in their activities of interest. For example, as a culminating activity for a unit in physical science that covers concepts in forces, motion, and simple machines, the teacher might task each student with building a representation of a machine designed to perform a specific task, such as taking out the trash or feeding the dog, that integrates some subset of the concepts covered.

Much like the Rube Goldberg cartoons of the past, these machines do not have to be practical. They can be made relevant and also conceptually correct as students describe them not only in terms of what they do but also in terms of how the concepts are present and what they are doing in the machines. In this manner, the teacher not only sees how the students are

The skateboard's upturned kick acts as a lever for the rider and helps lessen the force the rider exerts while performing tricks on ramps, the street, or level ground. Shown: William H. Roberston.

constructing knowledge but also provides students with a way to explain and elaborate on their individual understandings (Glasersfeld, 2001).

The teacher can provide each student with a rubric for the product they construct that focuses on the ideas and content to ensure that the information is accurate and current, the ideas come mainly from primary sources, the student demonstrates knowledge and insight, the machine shows effective use of technology, and all information relates to the lesson's overall purpose. The categories can be graded from 1 to 5, a scale in which a 1 signifies great need for improvement, a 3 defines a good job, and a 5 reflects an outstanding effort. In other words, the use of such a Likert scale can link to grades A through F.

This score in turn becomes the evaluation, and by using a rubric, the teacher gives the students choice in their learning as well as options to use various presentation or construction-of-knowledge methods, including computer software, visual materials, written materials, presentations, and demonstrations. As such, with the use of the 5Es of Constructivism as a framework for each activity, the teacher and students can see how the beginnings put forward in the engagement and the investigative nature of the explorations can effectively lead into content and concept materials in the explanation and elaboration phases. In the final phase of evaluation, the students have opportunities to put forth their answers and reflect on the process by which they reached their understandings. The entire process using the 5Es approach maps to each student's learning style, individual interests, and the curriculum needed to learn and master school.

SUMMARY

The use of action science as a mechanism for integrating transformative education is an approach that appears to be enhancing the interest and motivation of middle school science students. It is the purpose of *Action Science: Relevant Teaching and Active Learning* to positively impact achievement for middle school students in physical science knowledge and skills. By being immersed in a science learning approach based on action sports and focused on the goals and objectives of physical science, students' process skills and overall content knowledge could greatly increase. Studies have shown that students who are involved in active learning in meaningful contexts acquire knowledge and become proficient in problem solving (Robertson, 2008). The long-term prospects of this research area will seek to determine how the implementation of curriculum approaches built around student interests such as skateboarding and BMX can impact student achievement in science content and conceptual understandings.

REFERENCES

Bailer-Jones, D. M. (2002). Scientists' thoughts on scientific models. *Perspectives on Science, 10*(3), 275–301.

Davidson, C. N. (2011). *Now you see it: How the brain science of attention will transform the way we live, work, and learn.* New York, NY: Viking Penguin.

Dewey, J. (1902). *The child and the curriculum*. Chicago, IL: Chicago University Press.

Glasersfeld, E. V. (2001). The radical constructivist view of science. *Foundations of Science, 6*, 31–43.

Lee, O. (2002). Science inquiry for elementary students from diverse backgrounds. In W. G. Secada (Ed.), *Review of research in education* (Vol. 26, pp. 23–69). Washington, DC: American Educational Research Association.

National Research Council. (2006). *Learning to think spatially*. Washington, DC: National Academy of Sciences.

National Science Board. (2006). *America's pressing challenge—Building a stronger foundation: A companion to science and engineering indicators 2006*. Arlington, VA: National Science Foundation.

Robertson, W. H. (2008). *Developing problem-based curriculum: Unlocking student success utilizing critical thinking and inquiry*. Des Moines, IA: Kendall Hunt.

When you see riders going fast, you probably notice that they cover various distances in a short period of time. In other words, the force, distance, and time they cover while riding relates to their overall power. Shown: Daniel Dhers.

7 Action Science and the Future

Outliers are those who have been given opportunities—and who have had the strength and presence of mind to seize them.

—*Malcolm Gladwell (2008)*

In my work with K–12 students—with whom I have regular interactions due to the field-based nature of my work in the College of Education—I have come to see that students access information in new ways that differ from and extend those methods we usually use to deliver content to university students. For example, while many learners in my generation have grown up accessing textual information, the use of Google as an Internet search engine to find articles or relevant textual content is quite normal. Yet as discussed in the previous chapter, for those students currently in middle school, their first step is often to look for a YouTube video on a given topic and then, once they have seen and heard video content, to explore that topic through text and concrete experiences. This is the type of modern learner to whom I am appealing, as well as to the teachers working with these students.

To do this, I have developed a series of videos that combine real action performed by top action sports athletes with content delivery in physical science. The athletes in the video clips, which are 2 to 4 minutes long, execute high-flying maneuvers that demonstrate physical science concepts, such as the relationships between velocity and acceleration. Without the athletes, the action would not be as complete, and the action is another pathway that invites students to learn. They may not initially be attracted to science, but they might still recognize and respect the difficulty of the maneuvers performed in the videos. The videos provide participating teachers and students with a series of instructional opportunities and relevant content information that can be used to explore and explain the given content information, as well as to engage the students in classroom activities. Access to a number of these video clips, specifically to the ones cited in the activities in Chapter 8, is included as part of this book.

This presentation for K–12 teachers will focus on media and video instruction as a method of content dissemination in the areas of science, technology, engineering, and mathematics (STEM). Specifically, this area—dubbed "edutainment"—will chronicle three separate large-scale interactions with students, teachers, and community members that used live action and video as methods of capturing relevant content in student-centered contexts.

This type of engaging education connects to the methods that middle school students more commonly use, and this is an area with tremendous potential for outreach and informal education to inspire broader interest in science, mathematics, reading, learning, and other subjects. It can help develop a culture of education that extends democratically to make higher education accessible to all students (Singhal, Wang, & Rogers, 2012).

EDUTAINMENT AND ACTION SCIENCE

As we have discussed, the use of video instruction for middle school students can help motivate the exploration and explanation of physical science topics found in the areas of forces, motion, Newton's Laws of Motion, and simple machines. The use of familiar activities, situations, and objects—such as skateboarding and BMX—around which students can explore and explain scientific concepts can be defined as "action science." Action science is designed to teach fundamental physics concepts in an approach that uses transformative educational strategies, which help students move from memorizing facts and content to constructing knowledge in meaningful and useful ways. The activities in Chapter 8 associated with action science address the objectives and enduring knowledge of physical science in content and process skills aligned to national science standards.

Another idea is that such video content can be generated as a result of large-scale live demonstrations in multimedia-enhanced stadium settings, which can be used to engage students en masse. In turn, the produced video content can be used for in-class instruction and motivation in STEM-related topics.

For example, multiple such videos for action science have been created in conjunction with GEAR UP (Gaining Early Awareness and Readiness for Undergraduate Programs), a 5-year program funded by the Department of Education.

One such event was held at the Don Haskins Center at the University of Texas at El Paso on National GEAR UP Day (September 17, 2009). More than 3,500 eighth graders from the Ysleta Independent School District attended the event to learn more about basic scientific theories tested by several professional skateboarders, BMX riders, and an inline skater. The large-scale stadium event was developed to expand educational opportunities and assist students in becoming college eligible and academically successful in higher education. The Action Science Team used high-flying maneuvers to demonstrate physical science concepts such as the relationship between acceleration and velocity. The show incorporated action science concepts in the areas of forces, motion, and simple machines.

Momentum is another term that relates to motion, and skateboarders and BMX riders use momentum all the time. Momentum is the measure of mass in motion, and this property depends on both mass and velocity.

Building on the success of the previous effort, a second large-scale stadium demonstration was held for about 8,000 area elementary and middle school students in El Paso, Texas, in conjunction with Opportunity Nation festivities held on the University of Texas at El Paso campus on May 3, 2012. The team of professional BMX and skateboarding athletes performed a live demonstration to engage local students in explorations of mathematics and science in the context of action science. Additionally, a video project coordinated all the research, scripting, event footage, and interviews, as well as editing and postproduction, to create a final video product.

Another example of this type of edutainment comes from a unique way of delivering STEM content within a children's television show. In mid-September 2011, El Paso PBS affiliate KCOS-TV began airing a locally produced children's educational show called *Blast Beyond*, which features a seasoned television host, a three-piece rock band, and an on-stage audience of local schoolchildren ages 6 to 9. The show airs on weekdays to a broadcast area of 2 million people spanning three states, and past episodes are accessible on the KCOS-TV website (kcostv.org/localprograms/blastbeyond.php). Sample video of the educational aspects of the show are included in each archived and complete episode on the website and emphasize the types of interactions used within an educational television show designed to actively engage young students.

The basic format of the show consists of music, participation games, various skits, and a segment devoted to education, in which community educators act as content experts on academic topics. It is within this on-camera educational segment that educators provide an introduction for their lessons, as well as a 5-minute studio demonstration focused on an educational concept or specific content area. The show is in a "live-to-tape" format, which means that the cameras run continuously during a taping, with no editing except inclusion of various camera angles and title graphics. In a single day of taping, the show will be shot three times consecutively, with a short break between episodes. The total time for the three shows comes in at just under 2 hours, and this allowed a participating educator to be a guest on three shows per physical visit to the *Blast Beyond* set.

Overall, the impacts of generated real-life video instruction can be seen as one way of encouraging young people to explore higher education. This can be translated practically to the efforts of engaged educators who have appeared on the television show and designed active learning situations that explore diverse topics such as matter, balance, earthquakes, bugs, recycling, geometry, statistics, number sense, magnetism, and variables. The idea is that content can be presented in many memorable ways using a video format and video can be a viable resource to present fundamental content alongside textual information (Resnick, 1987).

THE TRULY FLIPPED CLASSROOM IS FUNDAMENTALLY CONSTRUCTIVIST

Another concept currently in vogue in education is the "flipped classroom," in which content is provided to the students outside of class and classroom time is spent primarily working on problems and hands-on activities. In theory, this is a huge step in the right direction, but without some real change in the fundamental method employed, it has the potential to revert back to the same old way of doing things. There is an expression that a pig in a dress is still a pig, and to call a strategy a "flipped classroom" while continuing to teach in a didactic, teacher-centered manner is not really flipped or progressive at all.

A truly flipped classroom, at its roots, has a constructivist approach embedded in the teaching and learning strategy. Fundamentally, in the current model of flipping the classroom, teachers often frontload the class with content and then make available in classroom settings educational opportunities that build on the content they provided. This is completely backward from the correct way to flip the classroom, which aligns at its core with the ideas of the 5Es of

Constructivism. The basic premise to be followed is not to provide content information first and then offer an experience based on the content; rather, the correct path is first to provide an experience based on the content and then to present the content information (Bybee et al., 2006).

As such, a simple way to flip the classroom is to start in the classroom, directing students to engage in activities in which they explore the content individually and in groups. For example, a lesson should begin with engagement and exploration, and in a flipped classroom design using the resources in this book, a video segment such as that on Newton's Second Law could be used to introduce the concepts and hook students into the lesson. The students should then be given time to explore the experiment titled "Force Makes a Mass Move" (see Activity 5 in Chapter 8) to gain a hands-on understanding of how force relates to mass with respect to acceleration. This is all done well before the teacher puts forward the definition of Newton's Second Law and prior to any delivery of specific content in terms of a lecture or similar didactic presentation.

Once the content has been explored in multiple ways and the discussions led by the teacher have been facilitative and filled with questioning strategies, the students should be given the primary content, whether it comes from a textbook, research articles, the Internet, or a video. The way you flip a classroom is to provide the experience first and the content second—in other words, to engage and explore in the classroom, explain and elaborate outside of the classroom, and then evaluate back in the class. The rhythm of this approach may seem easy enough, and the method does have a simple elegance, but the teacher must hold fast to the ideas of constructivism and keep the learner at the center, forgoing the urge to deliver content first and instead providing primary experiences first and content information second.

THE FUTURE FOR TECHNOLOGY, TEACHING, AND LEARNING IN ACTION SCIENCE

What will the future hold for learners now in middle school once they reach college and look to become part of the workforce? How will they engage with the content, and how will they learn? What pathways and mediums will become "normal," the standards by which learners access and consume content information? These questions and many more like them are guiding the discussions in teacher education and, in many ways, are setting us up for a future in which the learner has more access to content in more creative and interesting ways.

There is a push in higher education for more access to course materials, and many universities have responded to this by releasing massive open online courses (MOOCs) that are free and available to learners around the world. Once it is possible to get credit on college campuses for such courses, the enrollments on worldwide levels will likely skyrocket. In terms of action science, such courses are counterproductive, as the active learning focus is foremost in a learner-centered design. This is not to say that MOOCs or similar efforts could not adopt this medium—and, truly, use of computer-assisted instruction will only increase in the future—but to realize success within a constructivist approach, there will need to be a fundamental shift in the way the learner is positioned in the context of the learning experience.

MOOCs, in one sense, are the ultimate democratized educational vehicle, by which content, along with instruction, is offered to all free of charge—truly a pathway to providing education to all. But there is also something seemingly disconnected about the medium, and in its current form, it is really no more than a large-scale "sage on the stage" approach whereby rock star teachers deliver content to the masses. This is not a constructivist paradigm at all, and the student is not at the center of the learning; MOOCs revolve around the teacher as the bastion of knowledge.

Adaptive learning strategies using technology that provide direct feedback to students are also making great strides in education and have been used quite successfully in video games for many years. Many students, especially those digital natives in middle school today, enjoy playing video games, and the push toward "gamification" of educational content can be seen as one way video content can also be part of learner-centered and -directed video instruction. While this strategy does hold great potential for learning and is currently used in a number of vocations to train people to respond to situations through simulations, it has at its core the need for a guide, coach, and facilitator to offer students input on their performance and provide direction for next steps.

If you have ever watched a student play a video game, you understand that he or she will continue at the game until the goal is achieved. Should that goal be to reach the next level or to win a competition against others or to maneuver with greater speed through a maze or obstacle course, the learner is at the center of the activity and in control of the learning. In many ways, the idea of adaptive

A change in either the mass or velocity can result in a change in momentum. More mass at a similar velocity will create more momentum, as will more velocity for a constant mass. This can translate into bigger air on a ramp. Shown: Vic Murphy.

learning is the same, in that feedback is provided to learners in real time as they make choices in the content. While the results may not be the same as in a game, in which feedback for a bad choice may be complete annihilation, the method in many ways is the same: The learner gets direct responses based on choices made and connections established.

SUMMARY

The future of education is bright, and there will undoubtedly be a balance between technology, teaching, and learning as we move forward in the 21st century. The modern student will come to demand use of technology in the classroom, and the expectation is that students will be able to access content information in whatever form needed at any time, in any place. An instructor's use of action science content within constructivist pedagogy is at the heart of this text, and I hope it can be seen as a path for each learner to engage and explore content before ever experiencing a traditional lecture. The main purpose is to help teachers by providing a resource that is learner centered, with content a modern middle school student will find relevant and familiar. As students discover real-world connections within the classroom, it provides pathways to extend learning into the community, activating the many informal ways science content can be discovered, whether in museums, construction sites, or athletic competitions. Additionally, there is an underlying belief that using common objects and experiences as primary motivators for learning in action science and using the context of an activity based in youth culture, such as skateboarding, helps students be more interested and motivated to learn. Finally, this work is presented to appeal to teachers who have a heart for students who are disinterested in school and somewhat marginalized in the educational system, with the hope that these students can be found and inspired by educators who embrace the ideas in this book.

REFERENCES

Bybee, R. W., Taylor, J. A., Gardner, A., Van Scotter, P., Powell, J. C., Westbrook, A., & Landes, N. (2006). *The BSCS 5E instructional model: Origins, effectiveness, and applications; executive summary*. Retrieved from http://www.bscs.org/sites/default/files/_legacy/BSCS_5E_Instructional_Model-Executive_Summary_0.pdf

Gladwell, M. (2008). *Outliers: The story of success*. New York, NY: Little, Brown.

Resnick, M. (1987). *Edutainment? No thanks. I prefer playful learning*. Retrieved from http://llk.media.mit.edu/papers/edutainment.pdf

Singhal, A., Wang, H., & Rogers, E. M. (2012). The rising tide of entertainment-education in communication campaigns. In R. E. Rice & C. K. Atkin (Eds.), *Public communication campaigns* (4th ed., pp. 323–335). Thousand Oaks, CA: Sage.

The center of gravity, also known as the center of mass, allows for BMX riders to control both balanced and unbalanced forces in motion. Shown: Billy Gawrych.

8 Action Science Classroom Activities

ACTIVITY 1
FLATLAND BMX AND THE CENTER OF GRAVITY

ENGAGE

The first bicycle was reportedly built in the 1860s, and progress in the 1880s included pneumatic tires and better turning and handling. This type of bicycle allowed the rider to perform tricks and maneuvers, which soon appeared in circus and music hall performances. In the 1970s, bicycle "trick riders" began to advance a series of flatland maneuvers that were just starting to be invented. Flatland BMX is a form of cycling in which the rider spins, rolls, hops, and scuffs the tires while maneuvering the bike forward, backward, or on one wheel.

Billy Gawrych is a professional flatland BMX competitor and performer who travels across the country performing with GT Bikes and the Pro Impact Stunt Team. Billy typically performs a routine, often set to music, in which tricks are

69

linked together in a series of connected and flowing patterns. He is known for his high-speed tricks and for being able to perform maneuvers in front of many spectators. To perform such maneuvers, a rider such as Billy must control his or her center of gravity to balance and do cool tricks. In an individual, the center of gravity is a point that defines the center of his or her mass, and it is roughly located around the belly button.

Performers such as Billy add to the legacy of the sport and encourage BMX riders everywhere to try new tricks and seek out their own styles.

In addition, look at the action science video titled "Dr. Skateboard's Action Science—Forces—Episode 2—Center of Gravity," accessed with the QR code below.

EXPLORE

Purpose

The purpose of this activity is to show the student there is a relationship between force and motion. The student is expected to demonstrate how unbalanced forces, such as changes in the mass, size, or shape of an object, can cause changes in the center of gravity. In terms of the Next Generation Science Standards, this activity addresses the following task: Construct and present arguments using evidence to support the claim that gravitational interactions are attractive and depend on the masses of interacting objects.

Things You Need

- Plenty of string, cut in pieces of 20 to 30 cm in length
- Washers
- Thick card stock cut into 10-by-15 cm pieces (notecards will also work)
- Scissors
- Large paper clips
- Pencils with flat erasers
- Colored markers or colored pencils
- Hole punch
- Rulers

What to Do

1. The goal of this activity is to locate the center of gravity in a cardboard object that is not of uniform size or shape.

2. With a partner, take some card stock and draw an irregularly shaped object of your own design on the card stock. Your unique shape should not be a square, circle, rectangle, or any other symmetrical design. You should produce a creative, non-uniform shape.

3. Carefully cut out the shape you have drawn on the card stock.

4. Predict where your object's center of gravity is located, and mark it with a letter P. See if you can balance the object at the center of gravity you identified, using the eraser end of a pencil. Do not change your prediction.

5. Carefully punch two holes near any opposite edges of the cardboard structure, and use a colored marker or colored pencil to label the holes A and B.

6. Bend one end of a paper clip so it forms an L. Tie one end of the string to the paper clip and the other end to a washer.

7. Put the paper clip through the hole labeled A so that the object you created hangs freely from that point. Then, draw a line on the cardboard object that follows the vertical line of the string. Hang the cardboard object from the hole labeled B, and draw a second line that follows the vertical line of the string. Use a different-colored marker or pencil for each line.

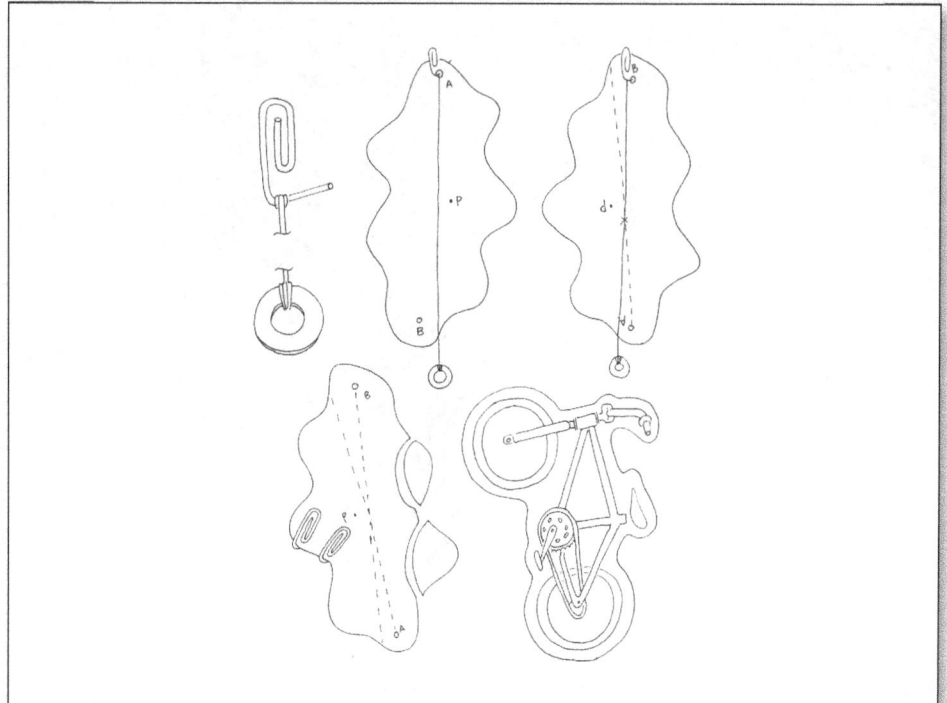

Figure 8.1

8. The center of gravity is the point at which the two lines intersect. You can test this out by trying to balance the object at this point on the eraser end of a pencil.

9. The main idea is that when an object is suspended at a single point, the center of gravity will hang directly below or at that point. To test this idea, you will use a piece of string attached to a washer to construct a vertical line beneath a point of suspension, and then choose another point and construct a second line. The center of gravity is where the two lines intersect.

10. Remove a piece of the object you have created and locate the center of gravity again. As the distribution of an object's mass changes, the position of the center of gravity also changes. Mark this new center of gravity with another line, in another color.

11. Finally, change the distribution of the object's weight by adding paper clips in such a way that the center of gravity moves back to the original location on the cardboard object.

12. As an extension of this exercise, try to construct a BMX bike out of the materials you used to make your previous shape, and see if you can determine the center of gravity of the BMX bike you created.

Placing the center of gravity over a point of balance can lead to new and exciting ways of riding a bike, such as balancing on the front wheel and handlebars. Shown: Billy Gawrych.

EXPLAIN

Because the cardboard cutouts the students made are irregular and asymmetrical, students should discover that the center of gravity is not necessarily found in the geometric center of an object. When students remove part of the cardboard cutouts they created, they will change its weight distribution. Students will then discover that the center of gravity has moved to a different point. Students should be encouraged to think of where the center of gravity is located in other objects, such as a boomerang, a basketball, or an empty cup. To move the center of gravity back to its original location on the cardboard, students need to add paper clips as weights near the area where they removed part of the cardboard.

ELABORATE

To be successful in skateboarding or BMX, a rider must understand how to control the distribution of his or her mass by controlling the center of gravity. In an individual, the center of gravity is a point that defines his or her center of mass, and it is located around a person's belly button. The closer the center of gravity is to the center of the board, bike, or rotational axis, the better chance the rider has of successfully completing a trick. Good riders tend to lower their center of gravity closer to the board or bike by bending their knees and shifting their weight.

Conversely, when a rider elevates his or her center of gravity by straightening his or her legs, the center of gravity extends too far from the board or bike and, inevitably, the force of gravity or centrifugal force overcomes the rider, leading to a crash. A rider's center of gravity needs to be over the center of the board or bike, and the best way to ensure that this happens is to bend the knees and lower the center of gravity. A person's center of gravity and the rider's ability to control the location of that center relative to the board or bike are the keys to progressing with bigger and better tricks.

EVALUATE

Once you have finished the procedures, answer these questions to draw some conclusions about what you have learned.

1. What is the center of gravity?

2. How does the center of gravity change as the size and/or shape of an object changes?

3. How can the center of gravity be altered?

4. How can the center of gravity be realigned in an object?

5. Where is the center of gravity located on a person?

Aligning the center of gravity over a pivot point and reducing the moment of inertia allows a rider to control the distribution of his or her mass in motion. Shown: Billy Gawrych.

ADDITIONAL TEACHER RESOURCES

Answers to Questions

1. *What is the center of gravity?* The center of gravity is a point that defines the center of an object's mass.

2. *How does the center of gravity change as the size and/or shape of an object changes?* The center of gravity moves if the change in the size and shape of an object is not uniform. For uniform changes, the center of gravity will stay in roughly the same place, such as when a person grows from a child to an adult. The relative position is still the same, but the exact location has changed.

3. *How can the center of gravity be altered?* Changing the mass of an object, the size of an object, or the shape of an object can change the center of gravity. You can affect one of these traits (or all of them) and alter the center of gravity.

4. *How can the center of gravity be realigned in an object?* Changing the mass of an object, the size of an object, or the shape of an object can realign the center of gravity. You can affect one of these traits (or all of them) and realign the center of gravity. In the experiment, the students changed the object's mass to realign the center of gravity back to its original position.

5. *Where is the center of gravity located on a person?* In a person, the center of gravity is located around the belly button.

Possible Extensions

Have students take different-sized objects of varying shapes and masses, and see if they can determine the center of gravity on each object. Students can bring in household items and use the same techniques in the lab to find the center of gravity. Objects could be empty soda cans, plastic jugs, books, Tupperware containers, or egg cartons, for example. Students should also be encouraged to make connections among an object's mass (and lack of mass), the distribution of the mass, and the location of the center of gravity. Finding this type of three-dimensional center of gravity will require the learner to look for new ways of determining it, such as trying to balance the items first. The student can also use the string approach but will have to use x-axis, y-axis, and z-axis intersections. As an extension, the teacher might pose the problem, "How can we determine a three-dimensional item's center of gravity and demonstrate that we have found it?" The students could use this open-ended exploration as a way to deepen and broaden their understanding of the center of gravity in real-life situations.

A rider's ability to use thrust to overcome forces of friction, or drag, can produce tricks that are exciting to watch and to perform. Shown: Daniel Dhers.

ENGAGE

"Go big or go home." That could be Daniel Dhers's underlying philosophy in riding BMX. Well-known and respected for his huge airs and massive transfers, Daniel goes higher, farther, and faster than most BMX riders today. He continually pushes himself to master new maneuvers as well as to improve standards in the sport of BMX. He is truly one of the most progressive BMX riders. He seemingly overcomes the forces of drag and gravity with a high degree of thrust and lift. In other words, the dude can go really fast and sky high.

Daniel hails from Caracas, Venezuela, yet competing in contests and performing demonstrations keep him constantly on the road. Even with all his success in contests, Daniel's attitude and efforts demonstrate that he is there primarily to have fun riding his BMX bike, since he truly has found a sport he

enjoys. An innovator in BMX, Daniel has accomplished feats on a bike that had previously never been attempted. He has perfected the backflip and countless variations, and is also known for his huge skate park transfers from ramp to ramp.

In addition, look at the action science video titled "Dr. Skateboard's Action Science—Forces—Episode 3—Thrust & Drag," accessed with the QR code on page 76.

EXPLORE

Purpose

The student knows that there is a relationship between force and motion. The student is expected to identify and describe the changes in position, direction of motion, and speed of an object when acted on by force. In terms of the Next Generation Science Standards, this activity addresses the following task: Construct and interpret graphical displays of data to describe the relationships of kinetic energy to the mass of an object and the speed of an object.

Things You Need

- Cardboard squares, 9 cm per side
- Thread spools
- Glue
- 9-inch balloons
- Scissors
- Lead fishing weights
- Masking tape
- Meter sticks
- Pencils

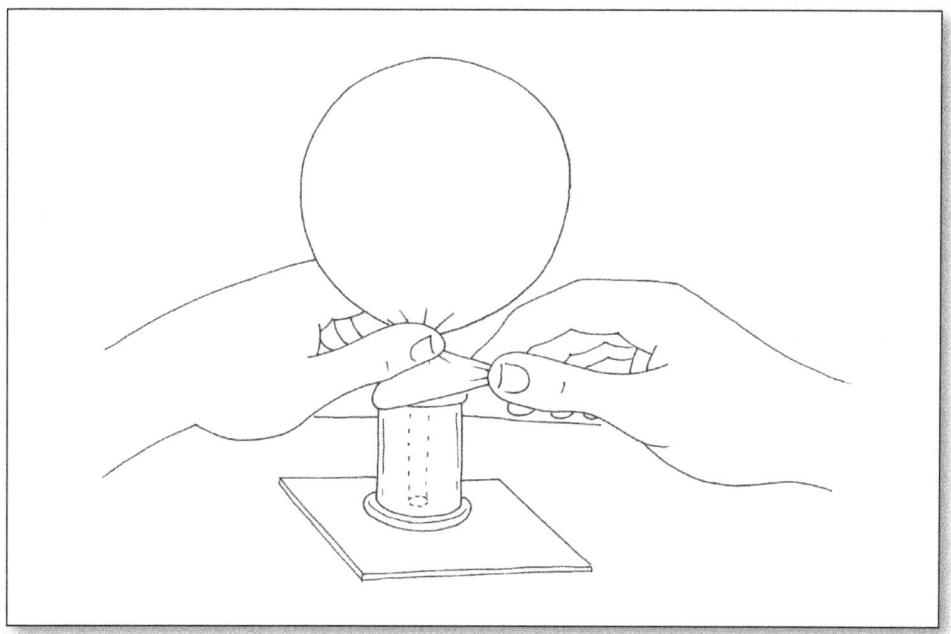

Figure 8.2a

What to Do

1. Punch a clean, round hole in the center of the cardboard square. You can use a hole punch or make a small guiding hole and work a pencil through the opening. The hole should have clean, smooth edges and should be the same size as the hole in the spool.

2. Glue the spool to the cardboard on top of the hole. Make sure you glue it really well and that the holes line up. Use enough glue to ensure that no air can escape between the spool and the piece of cardboard.

3. Cover the top of the spool with a circle of paper. Glue it to the spool, and let the glue fully dry.

4. Punch a hole in the middle of the paper cover where the hole of the spool is. Now your hole should run through the paper, spool, and cardboard without any obstructions.

5. Place a piece of masking tape on a flat surface. This is your starting line. Align your cardboard (with the spool) at the starting line. Give the cardboard and spool a little push with your hand.

6. Using the ruler, measure the distance between the starting line and the front edge of your cardboard. Record the distance on the data table.

7. Blow up the balloon and twist the end to keep the air from escaping. Measure the circumference of the inflated balloon. Next, stretch the balloon over the top of the spool.

8. Set the hovercraft on the starting line. Let go of the balloon, and give it a little push. Wait for all the air to escape the balloon, and then measure and record the distance.

9. Repeat Steps 7 and 8 at least two more times, and then average the results of the trials. Be sure to blow up the balloon to the same size for each trial.

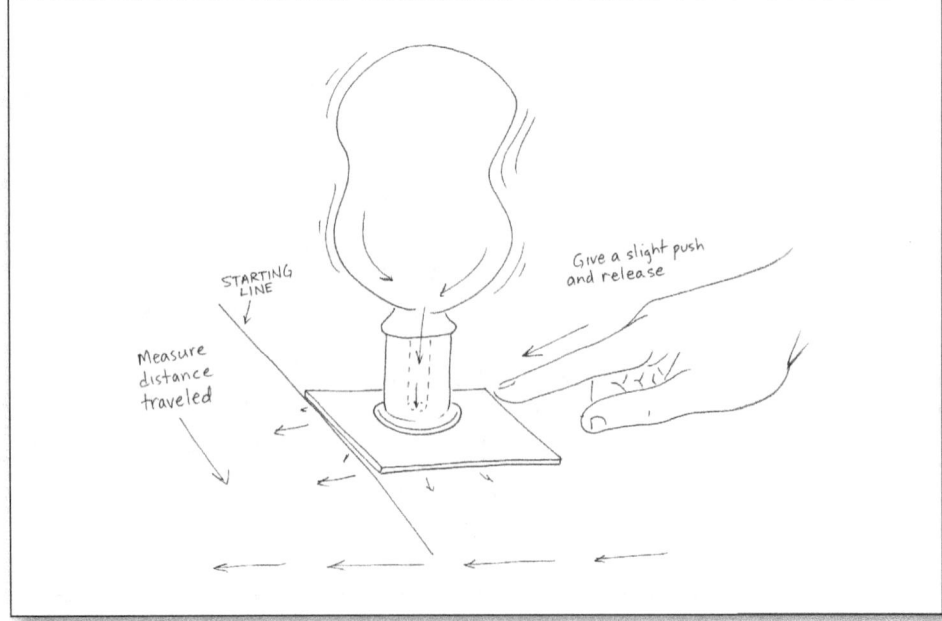

Figure 8.2b

One way to reduce drag is to fly through the air, which is a strategy BMX riders, as well as airplanes, use. Shown: Daniel Dhers.

10. Add lead fishing weights to your hovercraft, and repeat the trials. Be sure to record in your data table the weights you put on your hovercraft.

EXPLAIN

Please fill out the data tables below as you work through the "Explore" section. Record the distance traveled and circumference of the balloon in centimeters (cm). Record any weight added to the hovercraft in grams (gm).

Data Table 1

Trial	Distance Traveled (cm)	Circumference of Balloon (cm)	Weight Placed on Hovercraft (gm)	Percentage of Course Covered (%)
1				
2				
3				
Average				

Data Table 2

Trial	Distance Traveled (cm)	Circumference of Balloon (cm)	Weight Placed on Hovercraft (gm)	Percentage of Course Covered (%)
1				
2				
3				
Average				

Finally, calculate the percentage of the course covered by dividing the distance traveled by the total distance of the course and multiplying the result by 100.

In the case of a hovercraft, the air takes the shape of the bottom of the vehicle and the surface beneath it. In a real hovercraft, a skirt that forms a border around the bottom captures air under the vehicle. The idea is that the air forms a layer under the vehicle to reduce drag or friction, allowing the vehicle to move more easily. Real hovercrafts also have a fan that blows air under the bottom of the vehicle and keeps pushing out more and more air, thus increasing the pressure in the air cushion. The pressurized air cushion exerts a force on its container (the bottom of the hovercraft, the skirt, and the surface the hovercraft rests on). When the force exerted by this pressurized air grows to equal the weight of the hovercraft, it becomes buoyant (like a boat in water) and begins to float on air.

When a hovercraft travels over land or water, it acts a little differently than a boat or plane. For example, in the water, the pressurized air inside the air cushion pushes down on the water, causing some of the water to be displaced. Try this yourself. Blow into a sink full of water, and you will see that you create a small depression in the water. A modern hovercraft does the same thing, except it creates a larger depression.

ELABORATE

A hovercraft can fly on a cushion of air created by low-pressure, high-volume airflow at the base of the craft. An engine usually supplies this airflow. Another engine provides the thrust. There are also integrated systems in which one engine supplies the power for both lift and thrust. The air under the hovercraft is known as the air cushion. This air cushion leaks away under the bottom of the skirt to provide a film of air on which the hovercraft rides. The craft is steered by positioning a rudder in the thrust air stream to deflect the air and allow for changes in direction, much like a boat's rudder in water.

Hovercrafts fly on air and exert on the ground a pressure of less than a pound per square foot for most small machines. They are environmentally and ecologically friendly because they really just make a layer of air on the surface of the ground and use this air to fly. They are also quite easy to use and will provide a smooth and comfortable ride when the right balance of lift and thrust create a stable and forward-moving craft. Hovercrafts can be used to travel over grass, water, or sand. They are used by commercial enterprises for public transportation and by enthusiasts for sport. You can even buy a kit to build your own hovercraft.

EVALUATE

Questions to Answer

Once your experiment is done, see if you can answer these questions to draw some conclusions.

1. What characteristics allowed the original hovercraft to travel farthest?

2. How could a hovercraft be designed to minimize friction and maximize distance?

A BMX rider in flight manages the forces of gravity, lift, thrust, and drag to make a trick such as this no-handed aerial look easy. Shown: Daniel Dhers.

3. Which hovercraft was able to travel the farthest/fastest? Explain.

4. What is the effect of the air coming out of the balloon on the distance the hovercraft travels?

5. What happened as you increased the weight of the hovercraft? Explain.

6. Which conditions/characteristics (e.g., size of cardboard, balloon, holes, etc.) do you think are ideal for a hovercraft?

ADDITIONAL TEACHER RESOURCES

Answers to Questions

1. *What characteristics allowed the original hovercraft to travel farthest?* Typically, the hovercraft that travels the farthest has the cleanest hole in the cardboard that lines up with the spool. The hovercraft with sound construction that has secure interfaces between the cardboard/spool/balloon travels farthest. Each of these junctions can help or hinder student success. The idea is that the most efficient release of air from the balloon through the spool and under the cardboard is the key to success.

2. *How could a hovercraft be designed to minimize friction and maximize distance?* Minimizing friction could mean using a smoother ground surface to provide an even distribution of air as it exits the balloon. To maximize distance, the direction of the thrust needs to be efficient. This could be achieved by angling the hole or somehow directing the outflow of air in a generally backward direction.

3. *Which hovercraft was able to travel the farthest/fastest? Explain.* Teachers should focus on student-generated characteristics and use these responses to facilitate a list that incorporates previous answers.

4. *What is the effect of the air coming out of the balloon on the distance the hovercraft travels?* The more air that comes out quickly and efficiently, the greater distance the hovercraft will travel.

5. *What happened as you increased the weight of the hovercraft? Explain.* The added weight will cause the hovercraft to travel less distance if the balloon size stays the same. This may not be greatly observed with small increases in weight, but large increases in weight will cause a noticeable drop-off in distance traveled.

6. *Which conditions/characteristics (e.g., size of cardboard, balloon, holes, etc.) do you think are ideal for a hovercraft?* Teachers should focus on student-generated characteristics and use these responses to facilitate a list that incorporates previous answers.

Remind students that they can calculate the percentage of the total distance traveled by taking the measured distance traveled by their hovercraft (cm), dividing it by the overall distance (cm), and then multiplying the result by 100. For example, if distance traveled by the hovercraft is 75 cm and overall distance of the course is 100 cm, then the percentage of total distance traveled is 75 cm/100 cm x 100, which equals 75% of the course traveled.

Extensions

Students can engage in extension activities and compare outcomes for two hovercrafts. They can also create a hovercraft using a funnel and compare their outcomes to the hovercraft made with cardboard.

Students can determine the reasons for the variations in the outcomes in relation to the forces of thrust and drag on the systems. Students can also add weight to their models to increase the drag and make inferences about how well hovercrafts can sustain great amounts of weight.

Action science is one way of combining relevant activities, such as skateboarding, with discovery and engagement in science, mathematics, engineering, and technology. Shown: William H. Robertson.

ENGAGE

When Bill Robertson first became a university professor, he knew there would be plenty of opportunities to teach and serve within the community. With a PhD in science and technology education, Bill applies solid instructional principles in the college courses he teaches and, on a broader scale, in the community through outreach activities that feature skateboarding.

Bill performs skateboarding demonstrations with an added instructional twist on the topics of physics and mathematics. Using this platform to teach the physics of skateboarding has given him the unique identity of Dr. Skateboard. Bill says, "My audiences of children and parents typically do not see the connections between skateboarding and physics. They wonder, 'If you have a PhD, why do you skateboard?'" Bill says that through skateboarding he has learned patience, discipline, creativity, and the art and science of practice. His audiences include elementary, middle, and high school students in El Paso, Texas, and around the country.

Reference points allow someone on a bike or skateboard to determine if additional or less force is needed. A rider uses a point of reference to adjust his or her center of gravity, speed, momentum, or other physical factors in determining the speed or motion needed for a given maneuver or series of tricks. So when Bill is going for tricks on his skateboard, he uses points of reference on the ground, on a ramp, or in a park to manage the variables associated with perfecting maneuvers and tricks.

In addition, look at the action science video titled "Dr. Skateboard's Action Science—Motion—Episode 1—Fundamentals," accessed with the QR code below.

EXPLORE

Purpose

The student understands that motion cannot be determined unless a reference point is determined and that there is a relationship between force and motion. Also, the student is expected to demonstrate basic relationships between force and motion. The student is expected to identify and describe the changes in position as related to reference points, direction of motion, and speed of an object when acted on by force. In terms of the Next Generation Science Standards, this activity addresses the following task: Plan and conduct an investigation to provide evidence of the effects of balanced and unbalanced forces on the motion of an object.

Things You Need

- Plywood board (about 100 cm wide by 1,000 cm long)
- Textbooks or wooden blocks (four to six)
- Hot Wheels cars or fingerboards (one for each group of three students)
- Stopwatches (two)
- Meter sticks
- Masking tape

What to Do

1. Use a piece of masking tape to make a reference point about 15 cm from the top of the plywood board. This is the first reference point (PR 1).

2. Use the wooden blocks (or textbooks or something else) to construct a ramp out of the plywood. The ramp is an inclined plane that can also be considered a simple machine. Be sure to secure the plywood to the cinderblocks or hold it in a steady and consistent position.

3. First, place a piece of tape at the end of the plywood ramp. Record the distance (cm) from the first reference point (PR 1) on the data table. This line will be the second reference point (PR 2).

4. Second, place a piece of tape on the floor 40 cm from the end of the ramp. Record the distance (cm) from the first point of reference (PR 1). This line will be the third reference point (PR 3).

5. Measure the total distance from the starting line (PR 1) on the plywood to the third point of reference (PR 3) on the floor, and record the distance. This distance will be used to calculate the ratios of the distance covered by the cars or fingerboards within the points of reference.

6. Position a car or fingerboard at the first reference point (PR 1) at the top of the ramp, and let it roll down the ramp. Using the stopwatch, time how long it takes for the vehicle to pass the reference point at the end of the ramp (PR 2). Then time how long it takes to pass the reference point on the floor 40 cm from the end of the ramp (PR 3).

7. Try this for three separate trials and record the times (in seconds) and different points of reference on the data table. Calculate the averages of the ratios in the data table.

EXPLAIN

In the following experiment, you will need to calculate the percentage of the course covered by the fingerboard or car from the first point of reference (PR 1) to the second point of reference (PR 2) and then to the third point of reference (PR 3). You will also need to record the time in seconds the vehicle takes to cross

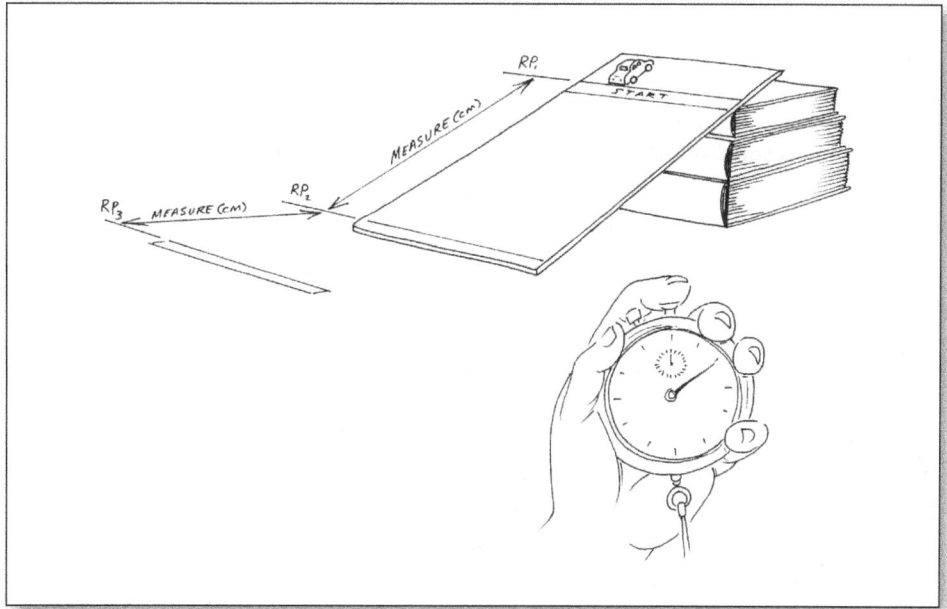

Figure 8.3

As a rider moves, motion can be detected and calculated based on the changes in distance from the point of reference over time. Shown: William H. Robertson.

each point of reference. Record these figures in the data table and also calculate averages. The following formulas should help.

$$\text{Total distance (cm)} = \text{Distance from the first reference}$$
$$\text{point (PR 1) to the third reference point (PR 3)}$$

$$\text{Percentage (\%) of course covered at PR 2} = ((\text{PR 2}-\text{PR 1})/$$
$$\text{total distance}) \times 100$$

$$\text{Percentage (\%) of course covered at PR 3} = ((\text{PR 3}-\text{PR 1})/$$
$$\text{total distance}) \times 100$$

Data Table

Trial	% of course covered at PR 2	% of course covered at PR 3	Time 1 for PR 2 (seconds)	Time 2 for PR 3 (seconds)
1				
2				
3				
Average				

Motion is defined as a change in position over time. The change can be in a horizontal direction, a vertical direction, or both. To detect motion, you need to have a reference point—in other words, something that is not moving or is moving at a different speed—to help relatively mark your movement.

When you're carving, turning, grinding, or going for air on your board or bike, you are experiencing motion. In skateboarding, you can use motion to help you go forward or backward. Going backward on a skateboard can be referred to as a "fakie" or "switch." In BMX, when the rider begins to head toward the ramp, the point of reference is where the rider begins. The cumulative change of position from the beginning to the end of a ride translates to the motion of the BMX rider.

ELABORATE

By pumping down the incline or through the point of transition, a rider uses the slope of a ramp or curve of a transition to provide an increase in the amount of force in the same direction. In action sports, when a rider pushes to get up a wall or down a ramp, the push is the force over a given distance resulting in work performed. Inclined planes are also used on highways as on-ramps, inclined planes that a car or motorcycle uses to gain speed. In contrast, the highway off-ramps are inclined planes that are designed to help a car or motorcycle slow down as it exits the highway.

EVALUATE

Once your experiment is done, answer these questions to draw some conclusions.

1. What is a point of reference?

2. How do you determine points of reference?

3. How does speed impact a point of reference?

4. How can reference points be used to calculate the distance an object travels?

5. How can reference points be used to calculate the speed at which an object travels?

Motion can be seen as the change in position from a point of reference, which can be achieved in creative ways on a skateboard, such as with this two-foot nose manual. Shown: William H. Robertson.

6. How is a reference point impacted by an inclined plane or the flat ground?

ADDITIONAL TEACHER RESOURCES

Answers to Questions

1. *What is a point of reference?* To detect motion, you need to have a point of reference—in other words, something that is not moving that can help relatively mark your movement.

2. *How do you determine points of reference?* You determine a point of reference by making a starting line from which to assess the motion. This allows an individual the opportunity to gauge a number of variables, including velocity, acceleration, and physical forces such as drag and thrust. The importance of a point of reference is that it provides a baseline from which to measure and impact a change in an object's motion.

3. *How does speed impact a point of reference?* The faster an object is moving, the less time it will have to be objective within a given point of reference. For example, a NASCAR racer can hit speeds of 140 mph on a curve, and the drivers often use various points on the track to determine if they need to brake or increase their speed to maintain control of the car. In action sports, a rider going for air off a ramp often uses multiple points of reference to determine the speed needed to negotiate the ramp and when to lift off the coping to perform an aerial.

4. *How can reference points be used to calculate the distance an object travels?* Reference points can be used as starting lines, finishing lines, or any other mark that stays fixed while a body is in motion.

5. *How can reference points be used to calculate the speed at which an object travels?* As an object moves through one reference point to another, the distance between the two points and the time it takes for the object to move between them can be used to determine velocity and speed.

6. *How is a reference point impacted by an inclined plane or the flat ground?* Since the point of reference is fixed, it is not impacted by an inclined plane or flat ground. It merely serves as a fixed spot from which to gauge changes in motion.

Extensions

Students can calculate velocity along with identifying points of reference. Remember that the initial velocity (Vi) is zero and the final velocity (Vf) is found using the calculation $V = d/t$, where d is the distance covered by the car or fingerboard and t is the time in seconds that it took to go down the entire length of the ramp. Students should do a minimum of three trials and use the data to calculate the velocity for each trial. Then they should determine an overall average velocity for their car or fingerboard's final velocity. Also, it is vital that they keep the units correctly associated with the formula and that velocity is a measure of how quickly an object moves or changes position.

ACTIVITY 4
ACCELERATION IS VELOCITY IN MOTION

Changes in velocity over time produce acceleration, which will increase as a rider moves down a ramp or inclined plane. Shown: Rayce Davis.

ENGAGE

Skateboarders riding half-pipes have a need for speed. The faster they go, the higher they can rise out of the pipe. Achieving greater heights is not only impressive but also necessary for pulling off acrobatic tricks such as Caballerials and McTwists. On flat ground, the conventional method for gaining speed is to push off with one foot. But half-pipes present a much more elegant option for the speed-hungry skater: pumping. To pump in a half-pipe, a skater first drops down into a crouch while traversing the flat bottom of the U-shaped pipe. Then, as the rider enters the sloped part of the ramp, called the transition, he or she straightens his or her legs and rises up. By raising his or her center of mass just at the beginning of the ramp's arc, the skater gains energy and thereby increases speed.

Pumping in a half-pipe is closely related to pumping on a swing. To get the swing to go higher, you lift your legs as you pass through the bottom of the swing's arc, then drop them at the top of the arc. Each time you do this, you gain a little energy and swing a little higher. From a physics point of view, the extra speed that comes from both kinds of pumping is a result of the rider moving between moments of acceleration and deceleration. As a rider moves into the bottom of the ramp, the moment of acceleration is greatest; as a rider ascends up the other wall, the forces of deceleration are enhanced. A successful rider can manage these forces to maximize the acceleration and minimize the deceleration, whether coming up or going down a wall.

In addition, look at the action science video titled "Dr. Skateboard's Action Science—Motion—Episode 3—Acceleration and Deceleration," located at the following URL:

EXPLORE

Purpose

The student knows that there is a relationship between force and motion. The student is expected to identify and describe the changes in position, direction of motion, and speed of an object when acted on by force. The student will demonstrate that changes in motion can be measured and graphically represented. The student can also demonstrate that an object will remain at rest or move at a constant speed and in a straight line if it is not being subjected to an unbalanced force. In terms of the Next Generation Science Standards, this activity addresses the following task: Construct, use, and present arguments to support the claim that when the kinetic energy of an object changes, energy is transferred to or from the object.

Things You Need

- Plywood board (about 100 cm wide by 1,000 cm long)
- Textbooks or wooden blocks (four to six)
- Hot Wheels cars or fingerboards (one for each pair of students)
- Stopwatches (two per group)
- Meter sticks
- Masking tape

What to Do

1. Use a piece of masking tape to make a starting line at the top of the plywood board.

2. Place the board on the textbooks or wooden blocks, creating an inclined plane. This ramp can also be considered a simple machine. Be sure to secure the board to the textbooks or wooden blocks to hold it in a steady and consistent position.

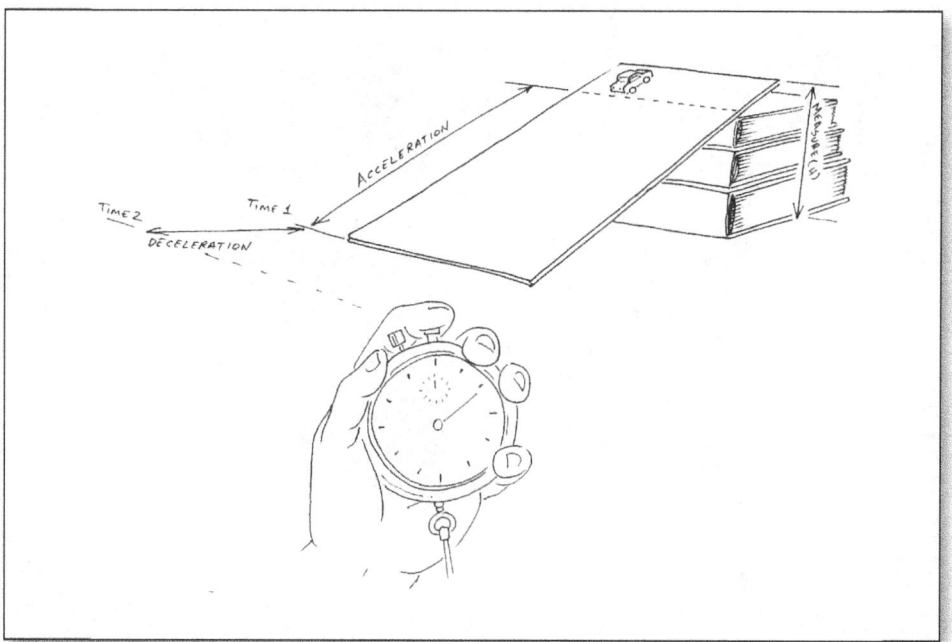

Figure 8.4

3. Measure the distance from the starting line to the end of the board, and record the distance. This number will be used to calculate the velocity of the cars or fingerboards.

4. Next, measure and record the height of the ramp from the floor to the top edge of the ramp.

5. Position your car or fingerboard at the starting line and let it roll down the ramp. Record the time it takes for your vehicle to descend and reach the bottom of the ramp.

6. Use the equation $V = d/t$ to calculate the velocity, and use this number as the final velocity (V_f).

7. Repeat the process, letting the car or fingerboard leave the ramp and continue over the flat surface or floor. You can measure the distance and record the time it takes for the vehicle to pass a certain point on the floor.

8. You can now use the equation $V = d/t$ to calculate the velocity from the end of the ramp to the point where the car or fingerboard stopped. This velocity and the time it takes for the vehicle to stop can be used to calculate deceleration.

EXPLAIN

For each set of trials, fill out each of the data tables. You will need to record the height of the ramp and distance traveled in centimeters (cm). You will also need to record the time (t) (in seconds) that it took the object to move over the distance traveled (d). Finally, you will calculate the final velocity (V_f) by dividing the distance traveled (d) by the time (t). Do this three times for each height from which you choose to test.

Performing an ollie, or no-handed aerial, is a standard way skateboarders move through space, and the use of acceleration can produce great results. Shown: Rayce Davis.

Data Table 1

Trial	Height of Ramp (cm)	d = Distance Traveled (cm)	t = Time (sec)	$V_f = d/t$
1				
2				
3				
Average				

Data Table 2

Trial	d = Distance (cm)	t = Time (sec)	$V_i = d/t$	$V_f = d/t$
1				
2				
3				
Average				

Data Table 3

Trial	d = Distance (cm)	t = Time (sec)	$V_i = d/t$	$V_f = d/t$
1				
2				
3				
Average				

Acceleration is a fundamental scientific concept that relates to motion. Subtracting the starting (or initial) velocity from the final velocity and dividing the result by the time calculates acceleration. Acceleration is a measurement of how an object's velocity changes in a certain amount of time. If the velocity is constant, there is no acceleration because there is no change in velocity. Deceleration is a special form of acceleration; it is defined as the loss of velocity over time, or negative acceleration as a moving object slows down. Both can be defined mathematically, provided you know the initial velocity, the final velocity, and the time it takes them to change. An acceleration equation that produces a negative answer is describing deceleration.

For example, a rider who drops in on a ramp has a velocity that changes from zero to a final velocity, which occurs at the bottom of the ramp. The acceleration is a calculation of how fast that change took place. If the rider had a final velocity of 2 m/s and it took 1.2 seconds to drop in on the ramp, can you calculate acceleration? In this example, the final velocity is 2 and the initial velocity is zero. Dividing the difference by 1.2 gives a final acceleration of 1.67 m/s/s, written as m/s^2.

The initial velocity (Vi) is zero on the ramp and the final velocity (Vf) is found using the calculation $V = d/t$, where d is the distance covered by the car or fingerboard and t is the time in seconds that it took the object to travel the entire length of the ramp. You will use the formula $a = (Vf—Vi)/t$ to calculate acceleration. Conversely, deceleration can be calculated by using the final velocity from the ramp and recording the distance and the time it took the car or fingerboard to stop. The students should do a minimum of three trials and use the data to calculate the velocity for each trial, and then determine an overall average number for their car or fingerboard's final velocity. Remind students that they must keep the units correctly associated with the formula and that velocity is a measure of how quickly an object moves or changes position.

A rider's ability to control motion by accelerating or decelerating is the key to accomplishing difficult maneuvers in motion. Shown: Rayce Davis.

ELABORATE

A rider uses the slope of a ramp or the curve of a transition to provide an increase in the amount of force in the same direction by pumping down the incline or through the point of transition. In action sports, when a rider pushes to get up a wall or down a ramp, the push is the force over a given distance resulting in work performed. Inclined planes are also used on highways as on-ramps, typically inclined planes that a car or motorcycle uses to gain speed. In contrast, the highway off-ramps are inclined planes that are designed to help a car or motorcycle slow down as they exit the highway.

EVALUATE

Questions to Answer

Once your experiment is done, see if you can answer these questions to draw some conclusions.

1. What is motion?

2. What is velocity?

3. What is speed?

4. How are velocity and speed similar? How are they different?

5. What is acceleration?

6. What is deceleration?

7. How does the height of an inclined plane impact an object's velocity?

ADDITIONAL TEACHER RESOURCES

Answers to Questions

1. *What is motion?* Motion is defined as a change in position over time. The change can be in a horizontal direction, a vertical direction, or in both directions.

2. *What is velocity?* Velocity is a measure of how quickly an object moves or changes position with reference to a particular direction, such as up and down or north and south. The positions of the units of velocity are the key to identifying it. For example, the distance is in centimeters and the time is in seconds, therefore the velocity will be a number with the units of cm/sec.

3. *What is speed?* Speed is a measure of how long it takes an object to move or change position. For example, the distance is in centimeters and the time is in seconds, therefore the speed will be a number with the units of

cm/sec. It is important for students to recognize the organization of the units in calculating and identifying speed.

4. *How are velocity and speed similar? How are they different?* Both measure a change in position over a specific time. They are different in that velocity inherently includes a direction, while speed does not.

5. *What is acceleration?* Acceleration is the rate at which the velocity of an object changes in a certain amount of time.

6. *What is deceleration?* Deceleration is a special form of acceleration. It is defined as the loss of velocity over time; so it is also the rate at which the velocity of an object changes over time, but it produces a negative result.

7. *How does the height of an inclined plane impact an object's velocity?* Typically, the steeper the inclined plane, the more velocity (or speed) an object can achieve. Students can verify this by doing different trials at different ramp heights and calculating the velocities for each case. It is important that they look for trends in the data they produce.

Extensions

For the experiment, the students can introduce another variable in the height of the ramp by using different numbers of wooden blocks or textbooks to establish different results. For example, students can use one block and run the experiment, add a second block and run the experiment, and do this all the way up to five blocks. The idea is to see the correlation between the height of the ramp and the changes in the object's velocity and acceleration. The teacher can also have students design their own vehicles out of classroom materials to see which designs work best, travel the greatest speeds, and cover the largest distances. The teacher can also lengthen the ramp, using other materials to increase the distance. This is also a good precursor to understanding the concepts of acceleration and deceleration. The different types of surfaces that cover the ramp can produce different results, as they will increase or decrease the drag. The main thing is avoiding the car or fingerboard coming to a complete stop so the student can get a calculation for the different velocities.

Formula for velocity: $V = d/t$
Formula for acceleration: $a = (V_f - V_i)/t$

The force a rider exerts in a trick, such as this BMX maneuver, requires the ability to make mass move in motion. Shown: Art Thomason.

ENGAGE

Spinning hitchhikers, 180-degree flips, tomahawk links, and breakers around the bike. These might sound like part of an overactive travel itinerary, but they are just some of the tricks performed by BMX professional flatlander Art Thomason. Art was born in Louisiana but moved to the small town of Ponca City, Oklahoma, as a child. Art has since taken up residence in Houston, Texas, where he works for the National Aeronautics and Space Administration (NASA). He also attended Texas A&M University, where he earned a master's degree in mechanical engineering to go along with his undergraduate degree in physics.

Art knows about hard work and practice. In fact, his work ethic in the sport is one of the things that set him apart from other riders. Art says, "In practice, I do every trick five times in a row, then do my run, and that usually helps me to pull my run perfectly." This has clearly worked well for Art, who has competed in four X-Games and two CFB Series, as well as getting a lot of national coverage on ESPN and ESPN2. Art now rides for one of the biggest names in the sport, Hoffman Bikes.

As a professional flatland BMX rider, Art believes that the large shows and national contests in which he participates have allowed him to interact with a lot of new riders and people interested in BMX. He relates the difference: "To the audience, the tricks look easy so they think they are easy. With the riders, it's really cool, because they know the difference." There is a lot of cooperation and camaraderie in the BMX flatland world. "The spirit of helping each other out, on a personal level, is really good. Although it can be competitive, everyone has a stake in it, so they help each other out."

In addition, look at the action science video titled "Dr. Skateboard's Action Science—Newton's Laws—Episode 2—Second Law," accessed with the QR code below.

EXPLORE

Purpose

The student knows there is a relationship between force and motion. The student is expected to demonstrate how unbalanced forces cause changes in the speed or direction of an object's motion. The student also must demonstrate that an object will remain at rest or move at a constant speed and in a straight line if it is not being subjected to an unbalanced force. The student is expected to identify and describe the changes in position, direction of motion, and speed of an object when acted on by force. In terms of the Next Generation Science Standards, this activity addresses the following task: Analyze data to support the claim that Newton's Second Law of Motion describes the mathematical relationship among the net force on a macroscopic object, its mass, and its acceleration.

Things You Need

- Small marbles
- Large marbles
- Meter sticks (at least three per group)
- Balance for weighing objects
- Masking tape
- Plywood board (about 100 cm wide by 1,000 cm long)

- Ring stand
- Iron ring
- Wooden blocks

What to Do

1. Looking at the different marbles you will use in the experiment, can you predict which will have greater forces? Make a series of predictions and record them in your notebook.

2. Select two marbles of different sizes and place them on a flat plane, such as the tabletop or the floor. Which marble do you think has more force? Why? If the marbles were at the top of an inclined plane, would they have potential or kinetic energy? Why? Record your predictions in your notebook.

3. Use the plywood board and either a ring stand and iron ring or wooden blocks to construct a ramp. The ramp should be about 15 cm high. This ramp is an inclined plane that can also be considered a simple machine. Be sure to secure the ramp to the iron ring or wooden blocks to hold it in a steady and consistent position.

4. Make a track by taping two meter sticks to the board, with a small gap between them for the marble to roll down the ramp.

5. Weigh each marble on the balance and record its mass on the data table.

6. At the top of the ramp, position one marble to drop in on the ramp. At the bottom of the inclined plane, position a second marble, either bigger or smaller than the marble at the top of the ramp. Hold the marble in place at the top of the incline.

7. Release the marble at the top of the ramp and try to hit the second marble. It may be useful to start at a lower distance on the ramp and work your way up to the top. Whichever you choose, be sure to use the same distance for all three trials in your experiment.

8. Measure the distance traveled by both marbles after the collision. Record the data on a piece of paper.

9. Repeat this process three times and calculate the average distance.

10. Once you have recorded the data on the small marble, switch the marbles' positions. Repeat Steps 6 through 9 three times and record the distances the marbles move. Average the results.

EXPLAIN

In the following data tables, record the mass in grams (g) of each of the two marbles, as well as the distance (cm) that Marble 1 traveled down the ramp. Additionally, record the distance (cm) that Marble 2 traveled after the collision with Marble 1.

The faster a rider moves, in terms of both velocity and acceleration, the greater force he or she can generate, which increases with increased mass. Shown: Art Thomason.

Data Table 1

Trial	Mass of Marble 1 (g)	Mass of Marble 2 (g)	Distance (cm) Traveled by Marble 1 Down Ramp	Distance (cm) Traveled by Marble 2 After Collision
1				
2				
3				
Average				

Data Table 2

Trial	Mass of Marble 1 (g)	Mass of Marble 2 (g)	Distance (cm) Traveled by Marble 1 Down Ramp	Distance (cm) Traveled by Marble 2 After Collision
1				
2				
3				
Average				

Newton's Second Law of Motion states that the acceleration of an object depends on the object's mass and the force applied to it. Since acceleration is a change in velocity with a change in time, both concepts can be explored in relation to Newton's Second Law. The important fact is that a force will cause a change in velocity; likewise, a change in velocity will generate a force. We have defined force as equal to mass times acceleration (F = ma), and now we can define acceleration as the force divided by the mass (a = F/m). Greater mass with the same force produces less acceleration, and a smaller mass with the same force produces more acceleration. Greater force on the same mass produces more acceleration, and less force on the same mass produces less acceleration.

ELABORATE

Have you ever been sitting at a traffic light and seen a large truck next to a small car? What happens when they both step on the gas and accelerate down the road? Often, the smaller car can accelerate quicker from a stop than the larger truck can, as the smaller car has less mass to accelerate. This will cause the smaller car to move ahead of the bigger truck initially. In other words, both the car and the truck can create equal forces even though they are of different masses and moving at different accelerations. If the car (smaller mass) has a high acceleration and the truck (larger mass) has a lower acceleration, they may be equal in force. In some ways, that is why a traffic accident can cause a lot of damage even at low speeds, as the force exerted by the cars is a product of both the mass and the acceleration of the cars.

In skateboarding and BMX, the constant combined mass of the rider and the board or bike defines a system that will go faster as more force is applied to it. As discussed in the forces segment, a skateboarder can maintain a constant-level

BMX is an important part of action science, and having fun in both activities and education is fundamental to learning new things. Shown: Art Thomason.

ollie, in which the force of gravity is balanced by the lift and the drag is balanced by the thrust. However, if the forces become unbalanced, the rider moves in the direction of the greater force.

EVALUATE

Questions to Answer

Once your experiment is done, answer these questions and draw some conclusions.

1. How do your predictions compare to the outcomes of the experiment?

2. Which marble has the most potential energy at the top of the ramp? Why?

3. Which marble has the most kinetic energy at the bottom of the ramp? Why?

4. What factors affect the amount of force the marbles can generate?

5. What would happen if the smaller marble were released and collided with a larger one?

6. What would happen if the larger marble were released and collided with a smaller one?

7. Predict what would happen if you used a tennis ball and a golf ball in this activity, and explain why you made that prediction.

ADDITIONAL TEACHER RESOURCES

Answers to Questions

1. *How do your predictions compare to the outcomes of the experiment?* Answers will vary, but the students should write their predictions and their reasons for these predictions. Remember to answer the students with questions, not provide answers. Appropriate facilitative questions might include the following: What do you think will happen? Why do you think that would be?

2. *Which marble has the most potential energy at the top of the ramp? Why?* Potential energy is defined as the energy of position. The larger marble at the highest point will have a greater potential energy (PE) and can be calculated as PE = mass x gravity x height. The force or potential energy will be tied to both mass and acceleration of the marble. The larger marble has more mass than the smaller marble, and since marbles moving down a similar inclined plane have a relatively equal acceleration, the force will be more a product of mass.

3. *Which marble has the most kinetic energy at the bottom of the ramp? Why?* Kinetic energy is defined as energy in motion. The force or kinetic energy will be tied to both mass and acceleration of the marble, and in that way, the heavier marbles released from higher positions on the inclined plane will have higher kinetic energy. The energy will not be completely kinetic until the marble reaches the bottom of the inclined plane.

4. *What factors affect the amount of force the marbles can generate?* Newton's Second Law of Motion (F = ma) states that the acceleration of an object depends on the mass of the object and the force applied to it. Since acceleration is a change in velocity with a change in time and on the ramp the acceleration of either marble is relatively constant, force will depend primarily on the mass of the marble.

5. *What would happen if the smaller marble were released and collided with a larger one?* The larger marble would move and would impact the movement of the smaller marble. The forces would be conserved, yet the greater mass would take more force to move compared with the smaller mass.

6. *What would happen if the larger marble were released and collided with a smaller one?* The larger marble would move the smaller one to a greater degree. The forces would be conserved, yet the greater mass would deliver more force to move compared with the smaller mass.

7. *Predict what would happen if you used a tennis ball and a golf ball in this activity, and explain why you made that prediction.* This is a nice extension to help students see that mass is not always tied to size or volume. Again, as an instructor, you should facilitate this and allow students to try these as extensions, record their observations, and report their findings.

Extensions

You could repeat this activity with added mathematics. Students can use a stopwatch to time how long it takes the marbles to reach the end of the inclined plane. With the data collected, the students can create a table composed of the time the marbles take to move and the distance they travel. With this information, they can determine the velocity and acceleration of the marbles and also calculate the force based on the equation F = ma.

A tabletop air shows the action–reaction in that the bike is positioned on one side and the rider on the other as they fly through the air. Shown: Vic Murphy.

ENGAGE

To be a BMX rider who can catch big air, a person must possess a special set of skills and a disciplined mind-set. One such individual, Vic Murphy, has demonstrated his expertise as an extreme athlete for many years, with hundreds of demos on street courses, half-pipes, and transition-laden ramps. He has consistently performed at many events, dropping in and hitting ramps, soaring above the crowd with tabletop airs, barspins, and tailwhips. This is quite impressive on a quarter pipe and can also be extended to the newest craze, the Mega Ramp.

On the Mega Ramp, both skateboarders and BMX riders speed down a 60- or 80-foot roll-in and fly over a 50- or 70-foot gap. The riders then land on the other side and are propelled down the landing area and ascend a 27-foot quarter pipe, from which they vault 20 feet above the top of the lip. The riders can reach speeds up to 40 mph, just on the roll-in, and then exceed that as they negotiate the landing on both the initial jump and the aerial on the quarter pipe.

In addition, look at the action science video titled "Dr. Skateboard's Action Science—Newton's Laws—Episode 3—Third Law," accessed with the QR code below.

103

EXPLORE

Purpose

The student knows there is a relationship between force and motion. The student is expected to identify and describe the changes in position, and the direction of motion of an object when acted on by force. The student should also be able to demonstrate that changes in motion can be measured and graphically represented. In terms of the Next Generation Science Standards, this activity addresses the following task: Conduct an investigation and evaluate the experimental design to provide evidence that fields exist between objects exerting forces on each other even though the objects are not in contact.

Things You Need

- String (50 cm)
- Scissors
- Plastic straws (one per person)
- Masking tape
- Meter sticks
- Balloons
- Markers
- Paper

What to Do

1. Blow up a balloon and let it go. What happens when you do that in class? In what direction does it travel?

2. Next, you will construct a rocket and use the balloon to see if you can get the rocket to fly in a straight line across the classroom.

3. Cut a 50-cm piece of string and mark the string at 1-cm intervals.

4. Insert the end of the plastic straw into the balloon.

5. Secure the balloon tightly to the straw using tape. There should be no gaps or ways for air to escape from the balloon except through the straw.

6. Blow up the balloon by blowing into the straw. Once you have gotten the balloon to the size you want, hold your finger over the end of the straw to keep the air from escaping.

7. Have a partner measure the circumference of the balloon, using the 50-cm piece of string you have marked at 1-cm intervals.

8. Establish a reference point in the classroom from which to fly all the rockets, and mark it with a piece of tape on the floor.

9. Let the rocket go by taking your finger off the straw. Record the distance it flies from the starting point. Repeat this procedure two more times for a total of three trials.

10. Repeat Steps 6 through 9, blowing up the balloon to three different sizes. Remember to record the balloon's circumference and the distance the balloon travels every time.

EXPLAIN

For each of the data tables, record the circumference of the balloon in centimeters (cm) and the distance traveled in each trial. Calculate the average distance traveled, and see if you can create three uniquely different averages for the balloon circumference to see if you can determine which circumference produces the largest average distance traveled.

From the side, you can see how the forces need to balance in terms of action and reaction so the rider can amplify the trick and land successfully on the other side. Shown: Vic Murphy.

Data Table 1

Trial	Balloon Circumference (cm)	Distance Traveled (cm)
1		
2		
3		
Average		

Data Table 2

Trial	Balloon Circumference (cm)	Distance Traveled (cm)
1		
2		
3		
Average		

Data Table 3

Trial	Balloon Circumference (cm)	Distance Traveled (cm)
1		
2		
3		
Average		

Newton's Third Law of Motion states that for every force or action there is an equal and opposite force or reaction. This law refers to two objects exerting forces in opposite directions, which happens every time a skateboarder or BMX rider takes a ride. In skateboarding, the rider exerts a push on the ground, force one, that is directed down and back as he or she moves toward the ramp. At the same time, the ground itself produces an equal force, force two, in the opposite direction. This causes the rider to move forward. So both the rider and the ground act as objects that produce forces in opposite directions.

ELABORATE

A number of action–reaction force pairs can be found in nature. One example is the way a fish moves through the water: The fish uses its fins to push water backward. At the same time, the water pushes the fish forward, propelling the fish through the water. In this case, the force on the water is directed backward and is opposite to the force on the fish, which is directed forward. In other words, the force on the water equals the force on the fish.

How about the way birds fly? A bird flaps its wings, which pushes air downward. At the same time, the air pushes back, lifting the bird upward. In this case, the force on the wings is directed downward and is opposite to the force on the air, which is directed upward. In other words, the force on the bird's wings equals the force on the air.

Let's take a look at something in BMX. As a rider is pedaling toward a ramp, what are the forces and what are the objects involved? In this case, the BMX rider (the first object) is pushing down on the pedals. This action produces a force (force one) as the rider moves toward the ramp. In what direction does this force push? At the same time, the ground itself (object two) produces an equal force (force two) in the opposite direction. In what direction does the force from the ground push?

EVALUATE

Questions to Answer

Once your experiment is done, answer these questions and draw some conclusions.

1. What is an action?

2. What is a reaction?

3. How are action and reaction related in Newton's Third Law of Motion?

4. What is the relationship between the circumference of the balloon and the distance traveled?

5. Which object is producing the action?

6. Which object is producing the reaction?

In terms of Newton's Third Law, BMX riders going for air, as in this photo, move between moments of potential and kinetic energy. Shown: Vic Murphy.

ADDITIONAL TEACHER RESOURCES

Answers to Questions

1. *What is an action?* An action is a force that moves an object in a given direction, such as forward, backward, up, or down.

2. *What is a reaction?* A reaction is a force that moves in the opposite direction of an action force.

3. *How are action and reaction related in Newton's Third Law of Motion?* "For every action, there is an equal and opposite reaction." On two interacting objects, such as a person sitting in a chair, there is a pair of forces interacting in opposite directions. The direction of the force on the person is opposite to the direction of the force on the chair. The size of the force on the person equals the size of the force on the chair.

4. *What is the relationship between the circumference of the balloon and the distance traveled?* Typically, the greater the circumference of the balloon, the farther it will travel.

5. *Which object is producing the action?* In the activity, the action is produced by the air moving out of the balloon. The air moves out of the balloon backward.

6. *Which object is producing the reaction?* In the activity, the reaction is produced by the balloon moving forward as the air moves backward out of the balloon.

Extensions

As an extension activity, have students play a game of tug-of-war in which you match teams of students in a contest to see which team can draw the other over the middle line. After a few trials and different teams, ask for a volunteer to take the rope on one side. Then ask for three volunteers to take the rope on the other side. Have students predict which side will win and provide reasoning for their prediction. The single student will generally lose the battle. Next, have the three students stand on a longboard (a skateboard about 4 feet long), and then have the students predict what will happen in the next tug-of-war. The class should then discuss the results and explore the reasons behind the results. Finally, have students report on how this activity is related to Newton's Third Law of equal and opposite reactions.

ACTIVITY 7
SKATEBOARDS HAVE LEVERS AND FULCRUMS

Simple machines are everywhere in skateboarding, including the inclined plane on this ramp and the levers, fulcrums, screws, wheels, and axles on the skateboard. Shown: Rayce Davis.

ENGAGE

In the 1950s, skateboarding started spreading across California. Surfers took to the streets to practice the moves they were performing in the ocean. The first boards were merely flat pieces of solid wood that took their shape and style from the surfboard. Modern skateboard decks have an upturned nose (or front) and upturned tail (or back). The upturned kicks act as a lever for the rider and help lessen the force the rider exerts while performing tricks on ramps, in the street, or on level ground.

The lever action of a skateboard allows a rider greater control of the board and makes tricks easier to perform. The lever on a skateboard allows the rider to perform ollies, nollies, tailslides, and blunts, to name just a few choice tricks. In skateboarding, the place where the trucks and the deck come together is an example of a fulcrum, or a fixed point around which a lever moves. The fulcrum allows the rider to control the movement of a trick by applying or releasing pressure on either side of the fulcrum.

In addition, look at the action science video titled "Dr. Skateboard's Action Science—Simple Machines—Episode 2—Lever & Fulcrum, Wheel & Axle," accessed with the QR code below.

EXPLORE

Purpose

The student knows there is a relationship between force and motion using simple machines such as levers and fulcrums. In terms of the Next Generation Science Standards, this activity addresses the following task: Develop a model to generate data for iterative testing and modification of a proposed object, tool, or process such that an optimal design can be achieved.

Things You Need

- Goggles
- Scissors
- Meter stick
- Tongue depressors
- Rubber bands
- Plastic spoons
- Plastic rulers
- Caps from water bottles
- Duct tape
- Masking tape
- Marshmallows (small and large)
- Modeling clay
- Balance for weighing objects

What to Do

1. Using duct tape, attach one of the tongue depressors to a plastic spoon so it extends the spoon's handle. Be sure to secure it tightly. This will support the catapult arm and form the primary lever.

2. Using duct tape, attach another tongue depressor to the lever. This will further extend the arm of your catapult.

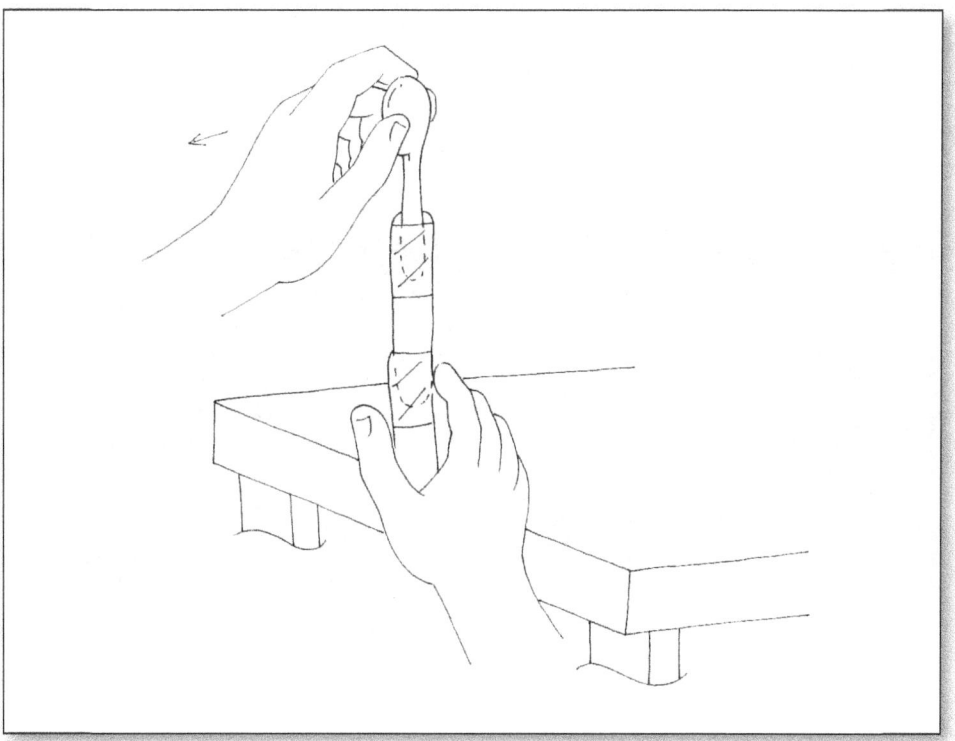

Figure 8.5a

3. Put your lever (or catapult) at the edge of the table and hold the end (or tongue depressors) tightly against the table. It is a good idea for one partner to hold the catapult and another to fire the objects.

4. Your catapult is ready for use! When you are using the catapult, always make sure to hold down the base (or tongue depressors).

5. Set a beginning reference point and mark it with masking tape. This will be your starting point to see how far your object flies.

6. Select a small marshmallow, weigh it, and record its weight in grams (g) on your data sheet.

7. With the catapult in place against the table, put a small marshmallow in the spoon and prepare to launch it.

8. When you have released the catapult, mark the final distance the object flew. Record this distance in centimeters (cm) on your data chart. Remember to calculate the actual distance of the object from its beginning to its final landing spot. A third member of the team should be marking the distance the object travels and measuring the distance flown.

9. Repeat the experiment two more times to complete a total of three trials for your experiment. Calculate the average flight of your object, and compare this with classmates' results.

10. Try another object such as a large marshmallow or piece of modeling clay and see what results you get. You can also make objects out of the modeling clay that can be of unique sizes and shapes. Be sure to record

all your data and look for similarities and differences among your catapults and those of your classmates.

11. As an extension, see if you can create a better catapult using the materials provided, including the plastic rulers and caps from water bottles.

The front and back of the skateboard are both upturned to function as levers over the trucks, which act as fulcrums. Shown: William H. Robertson.

EXPLAIN

For each object, describe it, weigh it, and record the distance it travels. Do this three times for each object you choose to launch.

Data Table 1

Trial	Name of Object	Weight of Object (g)	Distance Traveled (cm)
1			
2			
3			
Average			

Data Table 2

Trial	Name of Object	Weight of Object (g)	Distance Traveled (cm)
1			
2			
3			
Average			

Data Table 3

Trial	Name of Object	Weight of Object (g)	Distance Traveled (cm)
1			
2			
3			
Average			

Simple machines are devices that make work easier with a single motion and that consist of few or no moving parts. A machine is a device that does work, and simple machines make work easier. Although simple machines do not allow you to do less work, they do make the work easier in three main ways: (1) by lessening the force exerted, (2) by changing the distance over which a force is exerted, or (3) by changing the direction of the force exerted. Classic examples of simple machines are the screw, the wheel and axle, the wedge, the pulley, the inclined plane, and the lever. A lever is a straight rod or board that pivots on a point known as a fulcrum. Pushing down on one end of a lever results in the upward motion of the opposite end of the fulcrum.

One common purpose of levers is to enable a small force to overcome a large force. The lever is used for prying, as in the case of a crowbar, or for lifting. For example, the fulcrum is the point on which a crowbar rests when used to lift or pry loose some object; the effort is applied at the end farthest from the fulcrum and is relatively small. The distance from the operator's hands to the fulcrum is known as the lever arm, or effort arm; the object being pried loose is the resisting force, or resistance; the object's distance from the fulcrum is the resistance arm.

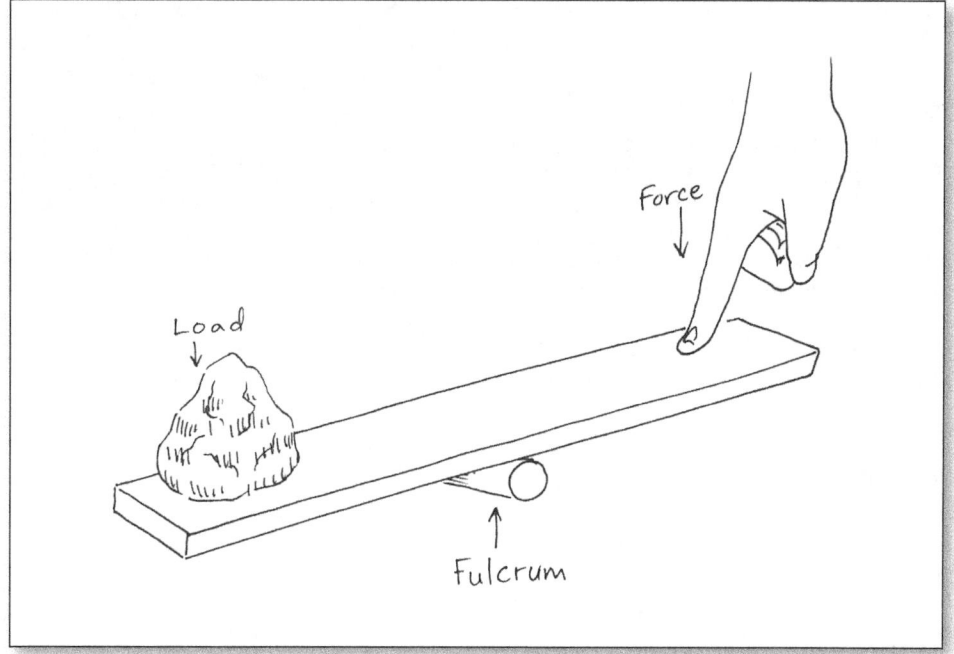

Figure 8.5b

ELABORATE

The modern skateboard has an upturned nose (or front) and tail (or back). The upturned kicks act as levers for the rider and help lessen the force the rider exerts while performing tricks on ramps, in the street, or on level ground. The lever action of a skateboard allows a rider greater control of the board and makes tricks easier to perform. It enables the rider to perform ollies, nollies, tailslides, and blunts, to name just a few choice tricks. In skateboarding, the place where the trucks and the deck come together is an example of a fulcrum, or a fixed point around which a lever moves. The fulcrum action allows the rider to control the movement of a trick by applying or releasing pressure on either side of the fulcrum.

Many other common tools, instruments, and appliances use the principle of the lever. Levers in which the fulcrum is located between the effort and the resistance, as in a crowbar and a double-pan balance, are known as first-class levers. The fulcrum may also be located at one end of the lever, with the effort applied at the other end and the resistance in between. This type of lever, illustrated by a wheelbarrow or nutcracker, is known as a second-class lever. The final possibility, known as a third-class lever, directs the effort between the fulcrum and the resistance. Examples of such a lever are tongs, brooms, fishing poles, and golf clubs.

The ollie allows a rider to fly, as when performing a high frontside air. In this maneuver, a rider pops the tail of the board and jumps, activating the lever action to increase distance and height. Shown: Lewis Dinsdale.

EVALUATE

Once your experiment is done, answer these questions and draw some conclusions.

1. What is a simple machine? How does it make work easier?

2. What is a lever?

3. What kinds of objects do you think of when you hear the word *lever?* List two levers that people use all the time.

4. What is a fulcrum?

5. What is force?

ADDITIONAL TEACHER RESOURCES

Answers to Questions

1. *What is a simple machine? How does it make work easier?* Simple machines are devices that make work easier with a single motion. Simple machines do not allow you to do less work, but they do make the work easier in three main ways: (1) by lessening the force exerted, (2) by changing the distance over which a force is exerted, or (3) by changing the direction of the force exerted.

2. *What is a lever?* A lever is a straight rod or board that pivots on a point known as a fulcrum. Pushing down on one end of a lever results in an upward motion at the opposite end of the lever.

3. *What kinds of objects do you think of when you hear the word lever? List two levers that people use all the time.* Answers will vary but will probably include objects such as see-saws or skateboards, or common tools such as crowbars or hammers.

4. *What is a fulcrum?* A fulcrum is a fixed point around which a rigid lever moves. The place where the trucks of the skateboard and the deck come together is an example of a fulcrum. The fulcrum action allows the rider to control the movement of a trick by applying or releasing pressure to the fulcrum point.

5. *What is force?* A force is any push or pull. It can be a helpful force, as in the way the wind pushes a sailboat forward on the water. It can also be a hindering force, such as when a salmon tries to swim against the current in a river.

Extensions

Math can be integrated into the lesson by having the students measure the distance the objects are catapulted. Participants may want to vary the objects they launch. Be sure that they will not injure other students as the objects are released. Marshmallows (used in this lab) are the safest option for this activity. From this experiment, participants may make a connection between mass of the object and the distance it is catapulted.

1. Identify some levers that are used in society.

2. What different classes of levers are found in the real world?

3. What is the relationship between force and distance?

4. How can force be measured?

Simple machines are devices that make work easier in a single motion. Although not all simple machines are used in skateboarding and BMX, a number are present. If you take two or more simple machines and put them together, you will get a complex machine. Both skateboards and BMX bikes are examples of complex machines. Can you identify all the simple machines found in a skateboard? Can you identify all the simple machines found in a BMX bike? How do you think the lever action of the skateboard helps riders perform tricks?

ACTIVITY 8
AROUND THE WORLD WITH WHEELS AND AXLES

Spinning in 360s is a way to balance centripetal and centrifugal forces while using the wheel-and-axle simple machine to keep the rotations progressing. Shown: William H. Robertson.

ENGAGE

Bill Robertson began skateboarding in 1976 and spent his time riding with his friends. They would challenge each other in everything they were learning: Who could ride the longest manual, do the fastest downhill, and complete the most 360 spins? Bill and his friends learned their tricks on the flat and then took them to the banks of ditches, ramps, and skate parks. Bill's abilities were developed with a freedom of expression in which he linked his tricks with personal style. As a skateboarder, Bill also entered contests, including slalom, high jump, park skating, half-pipes, pools, and freestyle. Freestyle, where riders deliver a series of their best moves in a routine integrated with music, is where Bill truly excelled.

In recent years, there has been an upswing in the "new school" and a nostalgic reflection on the "old school." The old-school approach reflects the roots of skateboarding, with an emphasis on moves that incorporate carving, rolling, turning, spinning, flipping, and most of all styling, which is performing tricks with style and freedom—in other words, freestyle. Bill continues to work and perform in skateboarding in a career that has spanned more than 30 years. He showcases freestyle skateboarding in demonstrations across the country and provides instruction in the art of freestyle skateboarding. There is such power, precision, and grace in freestyle, where a single rider performs all his or her best tricks in a choreographed routine on the flat ground. Freestyle is the essence of skateboarding, the roots of the boardwalk. Bill says, "Until someone refuses to get stoked from seeing me spin 20 1-footed 360s or throwing a handstand fingerflip, I will continue to embrace and promote freestyle."

In addition, look at the action science video titled "Dr. Skateboard's Action Science—Simple Machines—Episode 2—Lever & Fulcrum, Wheel & Axle," accessed with the QR code below.

EXPLORE

Purpose

The student knows there is a relationship between force and motion. The student is expected to demonstrate basic relationships between force and motion using simple machines such as pulleys and levers. The student is also expected to demonstrate that an object will remain at rest or move at a constant speed and in a straight line if it is not being subjected to an unbalanced force. In terms of the Next Generation Science Standards, this activity addresses the following task: Analyze data to support the claim that Newton's Second Law of Motion describes the mathematical relationship among the net force on a macroscopic object, its mass, and its acceleration.

Things You Need

- Plastic ruler
- Poster board or card stock
- Round pencils or 20-cm-long round dowels (four to six per group)
- Masking tape
- Plastic straws
- String (75 cm long)
- Markers
- Textbooks
- Spring scales of 250 g and 1,000 g

- Blocks of wood
- Compass to make circles of equal sizes
- Scissors

What to Do

1. Get some poster board or card stock, and cut out two circles about 10 cm or more in diameter. Next, use a pencil to punch or poke a hole in the center of each circle. The hole should be slightly smaller than the diameter of the straw. These structures will be the wheels for your device.

2. Insert the straw through both holes. This will be the axle for your device.

3. Roll the wheel and axle across the desk. The circles are the wheels, and the straw is the axle. If needed, you can secure the wheels to the axle with masking tape.

4. Analyze the design of the wheel and axle and record a list of machines that use the wheel and axle. Make observations and record them in your lab notebook.

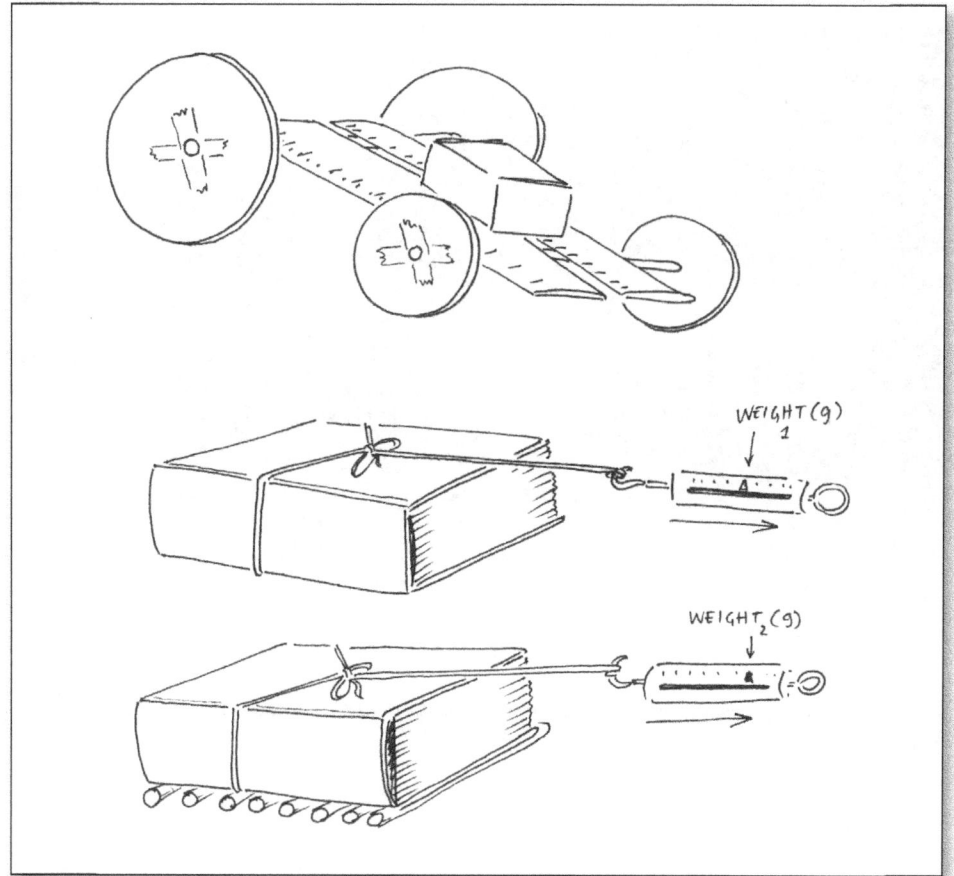

Figure 8.6

5. Next, make a second set of wheels and axle with smaller diameter wheels.

6. Compare and contrast the larger wheels and smaller wheels, and hypothesize the value of each setup. Observe what happens and record your observations in your lab notebook.

7. As a large group, brainstorm ways the wheels make work easier.

8. Get back into your small groups, and obtain a piece of string about 75 cm long.

9. Tie the string around a heavy book and drag it across the desk. Use the spring scales to record the force needed to move the book steadily across the desk.

10. Place a number of pencils or wooden dowels (four to six) under the book, and again pull it across the desk. Observe what happens and record your observations in your lab notebook.

11. Use your wheel-and-axle assemblies to move objects that fit between the axles, such as a small book, a block of wood, rulers, or other items around the classroom. Be sure to tape a plastic ruler to the axles so the object can rest easily on that surface. Observe what happens and record your observations in your lab notebook. Use the spring scales to record the forces.

A handstand is an example of using the wheels and axles of a skateboard in a creative and innovative way. Shown: William H. Robertson.

EXPLAIN

For each of the data tables, record in grams (g) the weight of each book or block of wood you use, the force without wheels, and the force with wheels.

Data Table 1

Trial	Weight of Book (g)	Force Without Wheels (g)	Force With Wheels (g)
1			
2			
3			
Average			

Data Table 2

Trial	Weight of Blocks (g)	Force Without Wheels (g)	Force With Wheels (g)
1			
2			
3			
Average			

Another simple machine to explore is the wheel and axle. The wheels and axles on a skateboard comprise the urethane wheels, the sealed bearings, and the axle that runs through the truck. There are wheels and axles almost everywhere we look. They are a very common and useful simple machine. On a skateboard, the wheel-and-axle simple machine allows the rider to roll, carve, grind, and spin.

In BMX, the bicycle wheel and the axle that runs through the center of the tire form the wheel-and-axle simple machine. The wheel-and-axle simple machines on both skateboards and BMX bikes allow riders to change the distance over which a force is exerted. In this case, the rider's input force is on the pedals and the output force comes from the tires' contact with the ground. Wheels and axles also make changing the direction of motion easier so that going forward or backward can happen quickly. The wheel and axle can be found in a number of modern technologies, including motorcycles, cars, buses, and airplanes. The wheels and axles on a BMX bike can also serve as fulcrum points for the frame of the bike to enhance maneuvers. In this case, the bike becomes a lever with the fulcrum on one wheel.

ELABORATE

A wheel-and-axle simple machine reduces the amount of friction an object creates in motion because fewer surfaces are exposed to the stationary object, usually the ground, at any given time. Examples of wheels and axles include tires, doorknobs, and the crankshafts on bicycles, steering wheels, gears, and egg beaters. The wheel and axle can be found in a number of modern technologies, including motorcycles, cars, buses, and airplanes.

Often, simple machines help reduce forces or gain leverage to make an action easier. They encompass the most basic of physics concepts and are found everywhere in our daily lives. We don't think about them as tools of science or ourselves as scientists when we use objects such as can openers, hammers, and dustpans, but these items act as simple machines in our daily lives.

Did you ever think of yourself as a physicist or engineer when you were riding your skateboard or BMX bike? In many ways, the traits that make a scientist are the same as those of a BMX rider or skateboarder. A physicist or engineer studies the interactions between concepts in physical science and attempts to create situations where the output forces are maximized by the input forces. Simple and complex machines help scientists maximize their input forces to make work easier. Skateboarders and BMX riders do the same thing, always trying to go higher and bigger on tricks.

With a skateboard, you can flip it from side to side and use the wheels not only for rolling but also as a platform on which to perform tricks. Shown: William H. Robertson.

EVALUATE

Once your experiment is done, see if you can answer these questions to draw some conclusions.

1. What is a wheel?

2. What is an axle?

3. Where are wheels and axles found in the real world?

4. How do you use wheels and axles in your daily life?

5. How do wheels and axles make work easier?

ADDITIONAL TEACHER RESOURCES

Answers to Questions

1. *What is a wheel?* A wheel is a solid disk or rigid ring connected by spokes to a hub, designed to turn around an axle passed through its center. In other words, a wheel-and-axle simple machine has a larger wheel (or wheels) connected by a smaller cylinder (axle) that is fastened to the wheel so they turn together.

2. *What is an axle?* An axle is a smaller cylinder fastened to the center of the wheel so it can turn with the wheel.

3. *Where are wheels and axles found in the real world?* Examples of wheels and axles include tires, doorknobs, and the crankshafts on bicycles, steering wheels, gears, and egg beaters. The wheel and axle can be found in a number of modern technologies, including motorcycles, cars, buses, and airplanes.

4. *How do you use wheels and axles in your daily life?* Answers will vary but will probably include objects such as cars, skateboards, bicycles, or some of the items from the question above.

5. *How do wheels and axles make work easier?* Wheel-and-axle simple machines help change the distance over which a force is exerted. Wheels and axles also make changing the direction of motion easier so that forward or backward motion can happen quickly. They also provide a distance advantage. Bicyclists put greater force on the pedals of the bike over a shorter distance to move the wheels a greater distance.

Extensions

The teacher can display pictures of Stonehenge and hypothesize how such large stones could have been moved or positioned in such a way in the past. The teacher can then ask students to use what they know about simple machines to determine how these huge slabs of stone were moved into place without the use of modern machinery. Students should be reminded that the rollers or logs they used to move their books are similar to the wheel and axle.

The teacher can discuss the issues of wheels and axles and ask students to write an essay about how the wheel has helped humans. Students can then make a list of the many uses of the wheel or let the students brainstorm as the teacher makes a list on the board or captures the results of the brainstorming activity with concept-mapping software. Students can then pick a topic from the list to build into a project and report on exactly how the wheel and axle is used to make work easier.

Index

CORWIN

A SAGE Company

The Corwin logo—a raven striding across an open book—represents the union of courage and learning. Corwin is committed to improving education for all learners by publishing books and other professional development resources for those serving the field of PreK–12 education. By providing practical, hands-on materials, Corwin continues to carry out the promise of its motto: **"Helping Educators Do Their Work Better."**

In compliance with GPSR, should you have any concerns about the safety of this product, please advise: International Associates Auditing & Certification Limited The Black Church, St Mary's Place, Dublin 7, D07 P4AX Ireland EUAR@ie.ia-net.com

www.ingramcontent.com/pod-product-compliance
Lightning Source LLC
Jackson TN
JSHW061923310126
97517JS00013B/127